Lecture Notes in Mathematics 1503

Salahoddin Shokranian

The Selberg-Arthur Trace Formula

Based on Lectures by James Arthur

Springer-Verlag
Berlin Heidelberg New York
London Paris Tokyo
Hong Kong Barcelona
Budapest

Author

Salahoddin Shokranian
Departamento de Matématica
Universidade de Brasília
70910 Brasília-DF, Brasil

Mathematics Subject Classification (1991): 22-XX

ISBN 3-540-55021-6 Springer-Verlag Berlin Heidelberg New York
ISBN 0-387-55021-6 Springer-Verlag New York Berlin Heidelberg

Printed in Germany

Typesetting: Camera ready by author
Printing and binding: Druckhaus Beltz, Hemsbach/Bergstr.
46/3140-543210 - Printed on acid-free paper

Introduction

Since the appearence of A. Selberg's famous paper in 1956, it gradually became clear that the role of Selberg's trace formula in automorphic forms, number theory, and representation theory is fundamental.

Selberg's paper is essentially concerned with the derivation of a trace formula for the regular representation of the group $PSL_2(\mathbb{R})$, and the application of this formula in calculating the trace of the Hecke operators for that group.

The importance of Selberg's paper naturally raised the question of how to generalize this trace formula to groups of higher ranks, and how to apply the formula in many problems directly or indirectly related to automorphic forms. Here we shall explain very briefly the attempts which have been made over many years to advance Selberg's theory of the trace formula.

Firstly, regarding the generalizations, a basic idea had already been indicated in Selberg's paper itself. This idea was that the Eisenstein series has an important role to play in advancing the theory from PSL_2 to the higher rank groups. One reason for this was the need to characterize the continuous spectrum of the regular representation of a Lie group G into the Hilbert space $L^2(\Gamma\backslash G)$, for a discrete subgroup Γ of G. Indeed, as a result of Langlands' work on the Eisenstein series, the continuous spectrum of the spectral decomposition was fully characterized and a major step toward the generalization of Selberg's trace formula was thus taken. On the other hand, the Duflo-Labesse paper in the late sixties, and J. Arthur's work in the early seventies were important, since the former provided thorough rigorous proof through representation theory of the work of Selberg, and the latter took the first steps toward the generalization of the trace formula for the higher rank groups. Finally, in his papers written since 1974 which begins by [1], Arthur has achieved the generalization of Selberg's trace formula for the arbitrary reductive group G, defined over a number field F, for the case where Γ is an arithmetic congruence subgroup of G. Arthur's works on the generalization of Selberg's trace formula are firmly based on three topics: the geometry of conjugacy classes in $G(F)$; the spectral theory of the Eisenstein series; and the theory of weighted orbital integrals. Arthur's trace formula thus establishes an identity between distributions obtained by conjugacy classes (termed the geometric expansion), and distributions obtained by the Eisenstein series (termed the spectral expansion). Moreover, this identity of distributions is valid simultaneously at all places of F, i.e., it is given globally for the adèle group $G(\mathbf{A})$ and its discrete subgroup of F-rational points $G(F)$.

Secondly, with reference to the applications, Selberg himself has shown in his paper and in other places that one may use his trace formula to obtain results of great importance in number theory and automorphic forms. Probably the most fascinating applications of Selberg's trace formula are now to be found in the works of Langlands, among which one should mention the Base Change theory of $GL(2)$, and also in algebraic geometry, which includes the theory of Shimura varieties, elliptic curves and corresponding zeta functions. One can always expect to obtain more general results using Arthur's generalized trace formula; for example, the Base Change theory of Langlands has recently been generalized to $GL(n)$ by Arthur and Clozel using the generalized trace formula.

This book is based on Arthur's lectures at University of Toronto during the fall term of 1987, which I had the opportunity to attend. Similar lectures were also given by Arthur during Jan. - Feb. of 1986 at the Tata Institute of Fundamental Research. My reasons for writing up the notes are two-fold. In the first place, I found the lectures very stimulating for grasping the basic ideas of the trace formula, and secondly, the trace formula is a fundamental tool in the study of many questions in automorphic forms and representation theory. Therefore, this book may serve as a basic reference for the fundaments of the important theory developed by Arthur.

I was first introduced to Selberg's trace formula by Prof. I. Satake ([85], [100]). As a curious student in Selberg-Arthur theory, I wanted to learn more. The result is this book.

This book consists of eight chapters. Each chapter has a brief introduction. There are a few exercises which are presented informally. They are basically intended to summarize the proofs and to help the reader to observe the more fundamental material which lies behind the theory of trace formulas, or to show the application of the trace formulas.

I would like to render thanks to Professor R. P. Langlands for introducing me to the works of Arthur, while I was a member of the Institute for Advanced Study during the 1986-87 academic year. I would also like to thank James Arthur for giving me the chance to visit University of Toronto and learn from him some of the most advanced mathematics of our time. His readiness and willingness to answer my questions and his friendship were vitally important.

I would like to thank the Brazilian Research Council (CNPq), Universidade de Brasília, the Institute for Advanced Study, the Canadian Research Council (NSERC), and University of Toronto for financial support. Finally, I would like to say that any inaccuracies which may remain are entirely my responsibility.

Salahoddin Shokranian
Brasília, July 1991

Contents

Chapter I

Number Theory And Automorphic Representations

In this chapter our aim is to explain how some basic issues in classical number theory, and modular forms, can be studied by means of representation theory of groups, and how some problems and conjectures in modern number theory are reinterpreted, or generalized by the use of automorphic representations. Thus, we hope that this expository chapter will serve as a motivating source for the further studies of automorphic representations, and the trace formulas.

1.1. Some problems in classical number theory

The problems to be considered in this section are of classical nature; they are related to the question of representation of a prime number as a value of a polynomial. More precisely, we explicitly state the following problem.

Problem A. For any polynomial $f(x)$ with integer coefficients, characterize the set of primes it represents.

Let $f(x) \in \mathbb{Z}[x]$ be a monic polynomial of degree r, we know that $f(m) \equiv 0 (\operatorname{mod} p)$ if and only if, $(x-m) \mid f(x) \,(\operatorname{mod} p)$. Thus, if a prime p is represented by $f(x)$ for the argument $x = m$, we can factor $f(x)$ into the product of irreducible polynomials $f_i(x)$ each one of respective degree $r_i (\operatorname{mod} p)$, $\quad (1 \leq i \leq \ell)$. But, we must have

$$r = r_1 + r_2 + \ldots + r_\ell.$$

Thus, the prime number p gives rise to a *partition* $(r_1, r_2, \ldots, r_\ell)$ of r. Now, we may conversely ask:

Problem B. Given a monic polynomial $f(x)$ of degree r, and a partition of r. For which prime p do we obtain this partition in the above manner?

This problem is related to the Galois theory of finite extensions of $\mathbb{Q}$. In fact, recall that by elementary Galois theory, $f(x)$ gives rise to a finite Galois extension $E/\mathbb{Q}$ by its splitting field, and this give rise to the Galois group $\operatorname{Gal}(E/\mathbb{Q})$ which is a finite subgroup of S_r (the symmetric group in r-letters). Observe that the partition $(r_1, \ldots, r_\ell)$ of r determines a conjugacy class in S_r, whose classes consists of permutations which decompose into disjoint cycles of lengths $r_1, r_2, \ldots, r_\ell$. Suppose that $\operatorname{Gal}(E/\mathbb{Q})'$ is the embedding of $\operatorname{Gal}(E/\mathbb{Q})$ into S_r. Then the intersection of $\operatorname{Gal}(E/\mathbb{Q})'$ with the conjugacy class of $(r_1, \ldots, r_\ell)$ in S_r may be several conjugacy classes in $\operatorname{Gal}(E/\mathbb{Q})$. Among these, there is *Frobenius class* which is distinguished by the condition that p does not divide the discriminant of $f(x)$. Hence, if p satisfies problem B, and does not divide the discriminant of $f(x)$ we may relate the problem above to the following one.

Problem C. Given a Galois extension $E/\mathbb{Q}$ and a conjugacy class X in $\mathrm{Gal}(E/\mathbb{Q})$, characterize those prime p for which X is the Frobenius class.

Recall that $E/\mathbb{Q}$ *splits completely* at p, or p is said to *splits completely* in E, if the Frobenius class of p is the identity in $\mathrm{Gal}(E/\mathbb{Q})$, i.e, if $f(x)$ breaks completely into linear factors mod p. Now, let $p(E)$ be the set of all primes at which E splits completely. Then, by Tchebotarev density theorem ([107], page 165) the map

$$T : E \to p(E)$$

is injective. This important result is used to state the final problem.

Problem D. What is the image of T in the map above (i.e, Tchebotarev theorem)? More precisely, what families of primes take the form $p(E)$?

As an example consider $f(x) = x^2 + 1$, $E = \mathbb{Q}(\sqrt{-1})$, $\mathrm{disc}(f(x)) = -4$. Then $p(E) = \{p : p \equiv 1 (\mathrm{mod}\, 4)\}$. For another example see ([30] page 354).

We now discuss the relation between the last problem and representation theory. In fact, the answer to Problem D is not an easy one. The problem has solution if $\mathrm{Gal}\,(E/\mathbb{Q})$ is abelian (as in the case of the example above). Then the solution is given by congruence relations (cf. [30] page 354). In the general case the solution of this problem is part of *non-abelian class field theory.* To explain it, one considers the inverse system of groups $\mathrm{Gal}(E/\mathbb{Q})$ of all finite Galois extensions E of $\mathbb{Q}$. Then, the inverse limit of this system is the compact totally disconnected topological group

$$\mathrm{Gal}(\bar{\mathbb{Q}}/\mathbb{Q}) = \varprojlim_{E} \mathrm{Gal}(E/\mathbb{Q}).$$

Now, one proceeds by considering a continuous homomorphism

$$j : \mathrm{Gal}(\bar{\mathbb{Q}}/\mathbb{Q}) \to GL(r, \mathbb{C}),$$

where continuity is with respect to the topological group $\mathrm{Gal}(\bar{\mathbb{Q}}/\mathbb{Q})$, which in particular means that j has finite image. On the other hand, since the kernel of j is equal $\mathrm{Gal}(E_j/\mathbb{Q})$ for some finite Galois extension E_j of $\mathbb{Q}$, we note that for any finite extension E of $\mathbb{Q}$, the Galois group $\mathrm{Gal}(E/\mathbb{Q})$ embedds into $GL(r, \mathbb{C})$. In fact, one can recover $\mathrm{Gal}(E/\mathbb{Q})$ as a quotient of $\mathrm{Gal}(\bar{\mathbb{Q}}/\mathbb{Q})$ with respect to a closed normal subgroup H of $\mathrm{Gal}(\bar{\mathbb{Q}}/\mathbb{Q})$. More precisely,

$$\mathrm{Gal}(E/\mathbb{Q}) \cong \mathrm{Gal}(\bar{\mathbb{Q}}/\mathbb{Q})/H.$$

Now, if p splits completely, the Frobenius class of $\mathrm{Gal}(E/\mathbb{Q})$ (which is identity) embedds into a unique semi-simple conjugacy class $\phi_p(j)$ in $GL(r, \mathbb{C})$. Hence we can write

$$p(E) = \{p : \phi_p(j) = I\}.$$

One can now relate $\phi_p(j)$ with certain L-functions. In fact, the standard procedure is to define an L-function by means of infinite product (Euler product) of *local L-functions*

$$L_p(s,j) = \det(I - s\phi_p(j))^{-1}, \qquad (s \in \mathbb{C}).$$

Then for a finite set S of primes, one defines the *L-function*

$$L_S(s,j) = \prod_{p \notin S} L_p(p^{-s}, j), \qquad (s \in \mathbb{C}).$$

The L-function above is related to the Artin's L-function. Concerning Artin's L-function, there is a famous conjecture (cf. [21], [51]), which asserts that the L-function which is originally defined only on some right half-plane, extends to an entire function on $\mathbb{C}$. This conjecture has not been proved in general, but our best knowledge about the validity of *Artin's conjecture* is the results that Langlands proved in his famous book [57]. We will return to this subject again in the last chapter of this book.

We now wish to explain the notion of an automorphic representation, and to mention some conjectures related to it.

1.2. Modular forms and automorphic representations

The concept of an automorphic L-function was introduced in the papers of Langlands ([58], [59]). Later, his ideas were extensively studied and developed by several mathematicians, in particular, Jaquet, Piateteski Shapiro, Shahidi, Shalika (cf. [44]). A classical approach, by which one can appreciate the notion of an automorphic representation, is to start with the concept of a modular form. Then, one has to realize a modular form as an automorphic representation. Therefore one can exploit the representation theoretical methods, and study certain problems in number theory. For this purpose one begins by recalling the notion of a modular form.

Let $\mathbb{H}$ be the Poincaré upper half plane, i.e, the set of all the complex numbers z with positive imaginary part. The group $SL_2(\mathbb{R})$ and its subgroups act on $\mathbb{H}$ by:

$$\begin{pmatrix} a & b \\ c & d \end{pmatrix} \cdot z = \frac{az+b}{cz+d}, \qquad (z \in \mathbb{H}).$$

Let N be a positive integer, and $\Gamma_0(N)$ the subgroup

$$\Gamma_0(N) = \{g \in SL_2(\mathbb{Z}) : \ c \equiv 0 (\mathrm{mod}\, N)\},$$

where $g = \begin{pmatrix} a & b \\ c & d \end{pmatrix}$. Defining $\mathbb{H}^* = \mathbb{H} \cup \mathbb{Q} \cup \{i\infty\}$, this space under certain topology becomes a Hausdorff, non-locally compact space [97]. Then, the action of $SL_2(\mathbb{Z})$ extends to $\mathcal{H}^*$, and the quotient space $\Gamma_0(N)\backslash\mathcal{H}^*$ turns into a compact Riemann surface. Now, a *holomorphic modular form of weight* $2k$, $(k \in \mathbb{Z}, k \geq 1)$, is a holomorphic function f on $\mathcal{H}^*$, that satisfies

$$f\left(\begin{pmatrix} a & b \\ c & d \end{pmatrix} \cdot z\right) = (cz+d)^{2k} f(z), \qquad (z \in \mathcal{H}^*).$$

Moreover, f is called a *cusp form of weight* $2k$ if it vanishes on the cusps $\mathbb{Q} \cup \{i\infty\}$. The holomorphic modular form f has a Fourier expansion at infinity of the form

$$f(z) = \sum_{n \geq 0} a_n e^{2\pi i n z}, \qquad (a_n \in \mathbb{C}).$$

Observe that, if f is a cusp form then $a_0 = 0$. We want here to recall the notion of a real analytic modular form, or Maass form. This is important for the statements of several conjectures, and also because of its historical connection with the work of Selberg on the trace formula.

Maass in his paper [73], considers a class of functions on $\mathcal{H}^*$ that satisfies the following conditions:

1. $f\left(\begin{pmatrix} a & b \\ c & d \end{pmatrix} \cdot z\right) = f(z)$, where $\begin{pmatrix} a & b \\ c & d \end{pmatrix} \in \Gamma_0(N)$,
2. $\Delta f = \frac{1}{4}(1 - s^2) f$, where $\Delta = -y^2\left(\frac{\partial^2}{\partial x^2} + \frac{\partial^2}{\partial y^2}\right)$ is the Laplace-Beltrami operator and $s \in i\mathbb{R}$, or s is a real number such that $-1 < s < 1$,
3. f is bounded,
4. f is a cusp form, i.e, f is zero on every cusp.

A function f, which satisfies the four conditions above is called a Maass cusp form of weight zero, or a *Maass form.* A Maass form has a Fourier expansion given by:

$$f(x + iy) = \sum_{n \neq 0} (|n|y)^{1/2} a_n K_{s/2}(2\pi |n| y) e^{2\pi i n x},$$

where K_ν satisfies the *Bessel* differential equation

$$z^2 \frac{d^2 K_\nu}{dz^2} + z\frac{dK_\nu}{dz} - (z^2 + \nu^2) K_\nu = 0,$$

and

$$K_\nu(z) \sim \sqrt{\frac{\pi}{2z}} e^{-z}$$

as $z \to +\infty$. For more details and discussions we refer the reader to ([65], [74]). It is now our objective to recall how modular forms or Maass forms can be studied in the context of automorphic representations. In fact, the space of these forms can be embedded into an L^2-space, and thus can be viewed as automorphic representations. The best way to explain this is to look at the adèlized group of $GL(2)$. However, let us first define the adèle group in a some what more general case, namely for $GL(r)$. Recall that, if A is any commutative ring with identity element 1, and $GL(r, A)$ is the set of all square matrices of size r with entries from A, and invertible determinant (i.e., invertible in A), then $GL(r, A)$ is a group under usual multiplication of matrices. Let,

$$A_1 = \mathbb{R} \times \mathbb{Q}_2 \times \ldots \times \mathbb{Q}_p, \ldots$$

be the product of $\mathbb{R}$ and the p-adic fields $\mathbb{Q}_p$ where p is a finite prime. Then, A_1 is a ring under component-wise addition and multiplication. Moreover, under the usual topology on $\mathbb{R}$, and the p-adic topology on $\mathbb{Q}_p$, A_1 is a topological ring. We are in fact interested in choosing a locally compact subring in A_1. For this, let $|x|_p$ be the p-adic absolute value of $x \in \mathbb{Q}^* = \mathbb{Q} - \{0\}$. As usual, denote by $\mathbb{Z}_p$ the ring of p-adic integers, $\mathbb{Z}_p = \{x \in \mathbb{Q}_p : |x|_p \leq 1\}$. Then $\mathbb{Z}_p$ is a compact ring in the topology of $\mathbb{Q}_p$ (indeed it is maximaly compact). Now, the ring of *adèles* is

$$\mathbf{A} = \{(x_\infty, x_2, \ldots) : x_p \in \mathbb{Q}_p, \text{ and } x_p \in \mathbb{Z}_p \text{ for almost all, but finitely many } p\}.$$

There is an obvious topology on A_1 (namely, *restricted direct product topology*), which makes $\mathbf{A}$ into a locally compact ring. We refer the reader for more details to [30]. It is now clear that $GL(r, \mathbf{A})$ is a locally compact group under the topology of $\mathbf{A}$. Observe that $\mathbb{Q}$ is embeded diagonally into $\mathbb{Q}_p$, and it is easy to see that under this embedding $\mathbb{Q}$ is a discrete subgroup of $\mathbf{A}$. In fact, choose the following neighborhood of zero in $\mathbf{A}$

$$\mathcal{N}(0) = \{\alpha \in \mathbf{A} : |\alpha_\infty|_\infty < 1, \quad |\alpha_p|_p \leq 1 \text{ for all finite prime } p\},$$

then $\mathcal{N}(0) \cap \mathbf{A} = \{0\}$. Similarly, $GL(r, \mathbb{Q})$ is diagonally embedded as a discrete subgroup into $GL(r, \mathbf{A})$. The fact that $GL(r, \mathbf{A})$ is locally compact, implies the existence of a right invariant measure on this group. This measure, in turn, induces an invariant measure on the quotient space $GL(r, \mathbb{Q}) \backslash GL(r, \mathbf{A})$. Observe that, the volume of this quotient space is not finite under this induced measure. However, if I is the identity matrix of size r, $\mathbf{A}^*$ the group of units of $\mathbf{A}$, and $Z(\mathbf{A}) = \{\lambda I : \lambda \in \mathbf{A}^*\}$, then

$$Z(\mathbf{A}) GL(r, \mathbb{Q}) \backslash GL(r, \mathbf{A})$$

has finite volume. At any rate, in both cases one can form the space of square integrable functions on the quotient space, which in the second case, it contains constant functions. Let us assume that

$$V = L^2(GL(r, \mathbb{Q}) \backslash GL(r, \mathbf{A})).$$

This is a Hilbert space which represents the vector space of a representation to be defined now. For any $y \in GL(r, \mathbf{A})$ and $\varphi \in V$, define

$$(R(y)\varphi)(x) = \varphi(xy), \qquad x \in GL(r, \mathbf{A}).$$

Then $R(y)$ is an operator on V, with norm

$$\begin{aligned} \|R(y)\varphi\|_2^2 &= \int_{GL(r,\mathbb{Q}) \backslash GL(r,\mathbf{A})} |\varphi(xy)|^2 dx \\ &= \int_{GL(r,\mathbb{Q}) \backslash GL(r,\mathbf{A})} |\varphi(x)|^2 dx \\ &= \|\varphi\|_2^2. \end{aligned}$$

Therefore, we have shown that $R(y)$ is a unitary operator. Moreover, under the usual composition of operators, R defines a unitary representation of $GL(r, \mathbf{A})$. This representation is called (*right*) *regular representation*. A general question which is related to the decomposition of representations, here also arises.

Problem 1. Decompose R explicitly into irreducible representations.

The problem above is fundamental in the theory of automorphic representations, and, in order to understand it, we must state some basic definitions. Suppose that G is a locally compact group, π a unitary representation of G into a Hilbert space H. Then π is *irreducible* if H has no closed subspace invariant under $\{\pi(y) : y \in G\}$, other than $\{0\}$ and H. We recall that, if π is not irreducible (i.e, *reducible*), and H_1 is a non-trivial invariant subspace, then $H_2 = H_1^{\perp}$, is also invariant under π, since π is unitary. It follows that, $\pi_1 = \pi|_{H_1}$ and $\pi_2 = \pi|_{H_2}$ are both unitary representations of G. In this case we write

$$\pi = \pi_1 \oplus \pi_2.$$

The following example is important because, it shows how the theory to be discussed in the next section, is related to harmonic analysis.

Example (1.1). Let $G = \mathbb{R}$, then G is a locally compact abelian group which has a regular representation into the Hilbert space $L^2(X)$, where $X = \mathbb{R}/\mathbb{Z}$. For any $y \in \mathbb{R}$,

$$(R(y)\varphi)(x) = \varphi(x+y).$$

On the other hand, since $\mathbb{R}$ is abelian, Schur's lemma implies that any irreducible unitary representation of $\mathbb{R}$ is one-dimensional. Using directly the definition of a unitary representation, and elementary facts from calculus, we see that the space of all unitary representations of $\mathbb{R}$, denoted by $\hat{\mathbb{R}}$ is parametrized by the real numbers, and it is given by

$$\hat{\mathbb{R}} = \left\{ \pi_\lambda : \pi_\lambda(y) = e^{i\lambda y}, \qquad y \in \mathbb{R}, \;\; \lambda \in \mathbb{R} \right\}.$$

Moreover, R decomposes completely into the direct sum of $\bigoplus_\lambda m_\lambda \pi_\lambda$, where the multiplicity m_λ is given by

$$m_\lambda = \begin{cases} 1 & \text{if } \lambda \in 2\pi\mathbb{Z} \\ 0 & \text{if } \lambda \notin 2\pi\mathbb{Z}. \end{cases}$$

Now, if we write

$$H_n = \{e^{2\pi i n\lambda} : \lambda \in \mathbb{R}\}_{\mathbb{C}}, \qquad (n \in \mathbb{Z}),$$

we have the following isomorphism

$$L^2(X) \cong \bigoplus_{n \in \mathbb{Z}} H_n$$

Moreover, if

$$(T\varphi)(n) = \int_0^1 \varphi(x) e^{-2\pi i n x}\, dx,$$

then T satisfies the following properties:

(i) $\|T\varphi\|_2 = \|\varphi\|_2$

(ii) $T \circ \pi(y) = \pi'(y) \circ T$,

i.e., T is an *intertwining operator*, between the two representations $\pi(y)$ and $\pi'(y)$ in $\hat{\mathbb{R}}$.

Let us now consider the case of $GL(2)$, and the Hilbert space $L^2(Z(\mathbf{A})GL(2,\mathbb{Q})\backslash GL(2,\mathbf{A}))$. First, recall that if

$$U_p(N) = \left\{ \begin{pmatrix} a & b \\ c & d \end{pmatrix} \in GL(2,\mathbb{Z}_p) : \;\; c \equiv 0 (\operatorname{mod} N) \right\},$$

then

$$GL(2,\mathbf{A}) = GL(2,\mathbb{Q})GL^+(2,\mathbb{R}) \prod_{p\nmid N} GL(2,\mathbb{Z}_p) \prod_{p|N} U_p(N),$$

where the right-hand side is a direct product of groups, and $GL^+(2,\mathbb{R})$ is the group of real 2×2 matrices of positive determinant. For a proof of this see [97]. Thus, an element $g \in GL(2,\mathbf{A})$ can be written as

$$g = \gamma g_\infty k_0 u,$$

where, $\gamma \in GL(2,\mathbb{Q})$, $\quad g_\infty \in GL^+(2,\mathbb{R})$, $\quad k_0 \in \prod_{p\nmid N} GL(2,\mathbb{Z}_p)$, and $u \in \prod_{p|N} U_p(N)$. Observe that one can define a *normalized automorphy factor* [97] for $GL^+(2,\mathbb{R})$ by setting

$$J(g,z) = (cz+d)(\det g)^{-\frac{1}{2}}.$$

Thus, since $GL^+(2,\mathbb{R})$ acts on $\mathbb{H}^*$ as before, one can similarly define the notion of a modular form or Maass form f. In particular, if we define

$$\delta_f(\gamma g_\infty k_0 u) = f(g_\infty \cdot i)J(g_\infty, i)^{-k},$$

(when f is a Maass form, $k = 0$), then, it can be shown [42] that

$$\delta_f \in L^2(Z(\mathbf{A})GL(2,\mathbb{Q})\backslash GL(2,\mathbf{A})).$$

Moreover, if f is a cusp form then,

$$\int_{\mathbb{Q}\backslash\mathbf{A}} \delta_f\left(\begin{pmatrix} 1 & x \\ 0 & 1 \end{pmatrix} g\right) dx = 0, \qquad (g \in GL(2,\mathbf{A})).$$

Thus,

$$f \to \delta_f$$

defines an injection of f into $L^2(Z(\mathbf{A})GL(2,\mathbb{Q})\backslash GL(2,\mathbf{A}))$.

The above injection justifies the term automorphic representation for the irreducible constituents of R. We now formally state the following definition.

Definition (1.1). π is called an *automorphic representation* of $GL(r,\mathbf{A})$ if it occurs in the decomposition of R.

In general, an irreducible unitary representation π of $GL(r,\mathbf{A})$ has a tensor product decomposition into local irreducible unitary representations π_∞ of $GL(r,\mathbb{R})$, and π_p of $GL(r,\mathbb{Q}_p)$ for all finite prime p. For a proof of this fact see [39]. We now give an example of a p-adic unitary representation.

Example (1.2). Let $B(\mathbb{Q}_p)$ be the group of upper triangular matrices in $GL(r,\mathbb{Q}_p)$. For an element $z = (z_1,\ldots,z_r) \in \mathbb{C}^r$, define

$$\chi_z(b) = |b_1|_p^{z_1+\frac{r-1}{2}} \cdot |b_2|_p^{z_2+\frac{r-3}{2}} \cdot \ldots \cdot |b_r|_p^{z_r-\frac{r-1}{2}},$$

where $b_1, \ldots, b_r$ are the diagonal elements of $B(\mathbb{Q}_p)$. Consider the space of all functions $\eta : GL(r, \mathbb{Q}_p) \to \mathbb{C}$ satisfying

$$\eta(bx) = \chi_z(b)\eta(x),$$

where $b \in B(\mathbb{Q}_p)$, $x \in GL(r, \mathbb{Q}_p)$. Then define,

$$(\pi_p(y)\eta)(x) = \eta(xy).$$

Thus, π_p defines a unitary representation of $GL(r, \mathbb{Q}_p)$ which is called the *unramified principal series representation.* We now give two properties of this representation. To this end, recall that the representations (π'_p, W'), and (π_p, W) are *unitarily equivalent*, if and only if, there is a norm preserving isomorphism $T : W \to W'$, such that

$$T \circ \pi_p(y) = \pi'_p(y) \circ T,$$

for all $y \in GL(r, \mathbb{Q}_p)$, T is then an intertwining operator. It is known that:

(i) If z is purely imaginary, π_p is both unitary and irreducible.

(ii) π'_p is unitarily equivalent to π_p if and only if

$$(z'_1, z'_2, \ldots z'_r) = (z_{\sigma(1)}, z_{\sigma(2)}, \ldots, z_{\sigma(r)}),$$

where $\sigma \in S_r$.

The unramified principal series representation is an important tool in the study of the automorphic representations of $GL(r, \mathbf{A})$. For example, if $\pi = \pi_\infty \otimes \pi_2 \otimes \ldots$ is an irreducible continuous representation of $GL(r, \mathbf{A})$, then almost all π_p are in the unramified principal series.

Now, define

$$\phi_p(\pi) = \begin{pmatrix} p^{-z_1} & & 0 \\ & \ddots & \\ 0 & & p^{-z_r} \end{pmatrix},$$

then one can regard it as a conjugacy class in $GL(r, \mathbb{C})$. Moreover, if S is a finite set of primes of $\mathbb{Q}$, and π is automorphic, then it is known that π is essentially determined uniquely by the family $\{\phi_p(\pi) : p \notin S\}$.

Let us state a basic conjecture towards the applications of an automorphic representation in Galois theory.

Conjecture 1. For any representation j of the Galois group $\mathrm{Gal}(\bar{\mathbb{Q}}/\mathbb{Q})$ into $GL(r, \mathbb{C})$, there exists an irreducible automorphic representation π of $GL(r, \mathbf{A})$ and a finite set S of primes, such that

$$\phi_p(\pi) = \phi_p(j) \qquad \forall\ p \notin S.$$

Example (1.3). Concerning the conjecture above, suppose that $r = 1$, that is, when the representation j is one-dimensional. Let $E/\mathbb{Q}$ be abelian. Then the conjecture is known, and it is the *Kronecker's theorem.* This theorem asserts that the field E is cyclotomic, i.e, E is contained in $\mathbb{Q}(e^{\frac{2\pi i}{N}})$, for some positive integer N. The proof is based on the notion of Hecke operator $T_{p,i}$. To define it, let

$$t_{p,i} = \mathrm{diag}(1, \ldots, 1, p, \ldots, p),$$

be a diagonal matrix of size r, such that the first $(r - i)$ elements are 1, and the others are all p. Let

$$K_p(N) = \{g \in GL(r, \mathbb{Z}_p) : g \equiv 1(\mathrm{mod}\, p^{\mathrm{ord}_p(N)} \mathbb{Z}_p)\},$$

where $\mathrm{ord}_p(N)$ is the highest power of prime divisor p of N. Moreover, suppose that $\chi_{p,i}$ is the characteristic function of the compact set $K_p(N) \cdot t_{p,i} \cdot K_p(N)$. Then, define the *Hecke operators* $T_{p,i}$ by:

$$T_{p,i}\varphi = \varphi * \chi_{p,i} = \int_{GL(r,\mathbb{Q}_p)} \chi_{p,i}(y)(R(y)\varphi)dy,$$

where $\varphi \in L^2(Z(\mathbf{A})GL(r, \mathbb{Q})\backslash GL(r, \mathbf{A})/K(N))$. As it is explained in [2], when $r = 1$, one uses Hecke operators $T_{p,0}$, and $T_{p,1}$ to show that the conjecture is true for any one-dimensional representation of $\mathrm{Gal}(\mathbb{Q}(e^{\frac{2\pi i}{N}})/\mathbb{Q})$, for some N. Hence, when $E/\mathbb{Q}$ is abelian, the character group of $\mathrm{Gal}(E/\mathbb{Q})$ is a subgroup of the character group of $\mathrm{Gal}(\mathbb{Q}(e^{\frac{2\pi i}{N}})/\mathbb{Q})$. Thus, $\mathrm{Gal}(E/\mathbb{Q})$ is a quotient of $\mathrm{Gal}(\mathbb{Q}(e^{\frac{2\pi i}{N}})/\mathbb{Q})$. This means that E is contained in $\mathbb{Q}(e^{\frac{2\pi i}{N}})$.

We now want to state the second conjecture, namely the functoriality conjecture. To explain it in full generality, one needs to appeal to the notion of *L-group* of a reductive algebraic group. The L-group of a reductive linear group is a complex subgroup of $GL(r, \mathbb{C})$. For example, the L-group of $GL(r)$ is $GL(r, \mathbb{C})$. This notion for the first time defined in [64], it is explained and exploited in [64]. However, since we want to keep this book as elementary as possible we prefer not to define the functoriality conjecture in terms of L-groups, for which the reader is refered to the papers [2], [64], [66], and [96]. We thus state *Langlands functoriality conjecture* for $GL(r)$. To this end, let us denote by $\mathcal{A}(GL(r, \mathbf{A}))$ the set of automorphic representations of $GL(r, \mathbf{A})$. As before, we denote by S a finite set of primes of $\mathbb{Q}$.

Conjecture 2. Let $GL(m_1, \mathbb{C})$, and $GL(m_2, \mathbb{C})$ be two general linear complex groups, and

$$\theta : GL(m_1, \mathbb{C}) \to GL(m_2, \mathbb{C})$$

be a complex analytic homomorphism. There exists a map

$$\begin{array}{rcccl} \theta' & : & \mathcal{A}(GL(m_1, \mathbf{A})) & \to & \mathcal{A}(GL(m_2, \mathbf{A})) \\ & & \pi & \to & \bigotimes_p \theta'(\pi_p), \end{array}$$

such that, if $\pi = \bigotimes_p \pi_p \in \mathcal{A}(GL(m_1, \mathbf{A}))$, then

$$\theta(\phi(\pi_p)) \subset \phi_p(\theta'(\pi_p)), \qquad \forall\, p \notin S.$$

This conjecture has been discussed and studied by several mathematicians, and it has been verified in certain cases, we refer the reader to the book of Gelbart and Shahidi [44], and to [59].

Example (1.4). A famous conjecture in number theory, and modular form is the so called *conjecture of Ramanujan and Petersson.* To describe it we do as follows. Say that a cusp form (holomorphic or Maass form) is normalized, if its first Fourier coefficient is $a_1 = 1$. Suppose that f is an eigenfunction for all the Hecke operators, then the conjecture asserts that the Fourier coefficients of the form at prime index p satisfy

$$|a_p| \leq \begin{cases} 2p^{\frac{k-1}{2}} & \text{if } f \text{ is holomorphic of weight } 2k, \\ 2p^{-\frac{1}{2}} & \text{if } f \text{ is a Maass form.} \end{cases}$$

We recall that when f is a holomorphic cusp form, the conjecture has been proved by Deligne [35]. It is interesting to note that, as it has been recently shown by Shahidi [96], Langlands conjecture of functoriality among other things, implies Ramanujan-Petersson conjecture.

Chapter II

Selberg's Trace Formula

This chapter consists of five sections. In the first one, we sketch some historical remarks. In the second, we derive Selberg's trace formula for a locally compact group G with respect to a discrete co-compact subgroup Γ. In the third section we give two examples, where we show how Selberg's trace formula generalizes two important theorem, one in analytic number theory, and the other in finite group theory. In the fourth section we show that the co-compactness condition on Γ is necessary for the convergence of the integral of Selberg's kernel function. Then, in the last section we discuss some generalization, and some applications of the trace formula, respectively.

2.1. Historical remarks

Suppose that G is a locally compact group and Γ a discrete subgroup of G. The existence of a left (or right) invariant measure on G induces a left (or right) invariant measure on the quotient space $\Gamma\backslash G$. Suppose that, under this induced measure the volume of $\Gamma\backslash G$ is finite, then Γ is called a *lattice* in G. Moreover, if the quotient space is compact, then Γ is called a *uniform lattice* or *discrete co-compact subgroup* of G. From the geometric point of view, and harmonic analysis, the spectral decomposition of the L^2-space $L^2(\Gamma\backslash G)$ is both, basic and a difficult problem. The spectral decomposition of this space in general consists of two invariant subspaces, a discrete part (discrete spectrum), and a continuous part (continuous spectrum). For example, if Γ is a uniform lattice in G, then the regular representation operator on $L^2(\Gamma\backslash G)$ has only discrete part. This case was actually considered by Selberg in his paper of 1956, whose interest, among other things was to characterize the continuous spectrum of certain integral operators acting on the L^2-space $L^2((SL(2,\mathbb{Z})\backslash SL(2,\mathbb{R}))$. As it is well known and it is easy to prove, $SL(2,\mathbb{Z})$ is a non-uniform lattice in $SL(2,\mathbb{R})$, i.e, the quotient space $SL(2,\mathbb{Z})\backslash SL(2,\mathbb{R})$ is non-compact and has finite volume. But, the geometry of this quotient space is relatively simple in the sense that, the *boundary components* (*cusps*) of $SL(2,\mathbb{Z})$ on the upper half-plane $\mathbb{H}$ are point-like, i.e, they are zero-dimensional. On the other hand, as a result of non-compactness, the spectral decomposition of the integral operators considered by Selberg on the space $L^2(SL(2,\mathbb{Z})\backslash SL(2,\mathbb{R}))$ has continuous spectrum, and Selberg's aim was to characterize this spectrum. In fact, this was one of the problems considered by him in his paper.

We now describe explicitly the operators considered by Selberg, and their relation with the Laplace-Beltrami operator Δ. Let $\gamma \in GL(2,\mathbb{Q})$, and $\Gamma = SL(2,\mathbb{Z})$, then $\Gamma^\gamma = \gamma^{-1}\Gamma\gamma \cap \Gamma$ is a subgroup of finite index in Γ. Thus, if γ also belongs to $GL^+(2,\mathbb{R})$, one can define the *Hecke operator*

$$T_\gamma : f \to \sum_{\delta \in \Gamma^\gamma\backslash\Gamma} f(\gamma\delta z), \qquad (z \in \mathbb{H}),$$

on the space of Maass forms. The basic observation is that; this operator commutes with Δ. Now, let $SO(2,\mathbb{R})$ be the orthogonal group of 2×2 real matrices. Then, a function φ on $\mathbb{H}$ as we have seen, may be identified with a function δ_φ on $SL(2,\mathbb{R})$ which is invariant on the right by $SO(2,\mathbb{R})$, through

$$\delta_\varphi\left(\begin{pmatrix} a & b \\ c & d \end{pmatrix}\right) = \varphi\left(\frac{ai+b}{ci+d}\right).$$

From harmonic analysis point of view, if f is a function with compact support on $SL(2,\mathbb{R})$, and *bi-invariant* (i.e, from left and right) under the action of $SO(2,\mathbb{R})$, then Selberg considers the operator

$$\varphi \stackrel{R(f)}{\rightarrow} \varphi * f,$$

where

$$(\varphi * f)(g) = \int_{SL(2,\mathbb{R})} \varphi(gx)f(x)dx.$$

In particular, these operators commute with each other and their spectral theory is identical with that of Δ. This is the class of operators studied by Selberg [93] (see also, [36], [67], [109], [110]).

The second problem considered in [93] was the calculation of the trace of $R(f)$ on the discrete spectrum, corresponding to a uniform lattice in $SL(2,\mathbb{R})$. This trace is expressed as a sum of orbital integrals, and it is known today as *Selberg's trace formula.* As an application, Selberg calculates the trace of Hecke operators acting on the space of cusp forms, using his trace formula. In his paper [93], Selberg mentions the importance of the study of Eisenstein series and their role in the characterization of the continuous spectrum of $R(f)$, for groups of higher dimensions. This program was carried out in full generality by Langlands in [68], (cf. [48]). Moreover, Selberg mentions that, $R(f)$ must be of trace class on the discrete spectrum. This was proved by Müller [83]. Implicitly, one asks for the generalization of Selberg's, trace formula, to a locally compact group with respect to a non-uniform lattice. This program was successfully carried out by Arthur, for reductive algebraic groups over a number field (cf. Arthur's papers since 1974).

2.2. Orbital integrals and Selberg's trace formula

After the above historical remarks we are now ready to write down Selberg's trace formula for a pair (G,Γ), where G is a locally compact group and Γ is a uniform lattice in G. Let us first recall that for certain locally compact groups, there always exists a uniform lattice. To begin, suppose that G is a *unimodular group* i.e, any left invariant measure on G is also right invariant. The class of unimodular groups is quite big and it includes familiar groups like $GL(r,\mathbb{R}), SL(r,\mathbb{R}), U(r,\mathbb{R}), SO(p,q,\mathbb{R})$. In general, the existence of a uniform lattice in these real groups is proved using a theorem of Borel [24], and even for the group $GL(r,\mathbb{Q}_p)$ one can deduce the existence of a uniform lattice from a theorem of [27], ([81], [86]).

Now, as an example observe that in $\mathbb{R}$ any discrete subgroup is of the form $\alpha\mathbb{Z}$, where $\alpha \in \mathbb{R}$. Then, the quotient space $\alpha\mathbb{Z}\backslash\mathbb{R}$ is compact, being homeomorphic to a circle. Thus $\alpha\mathbb{Z}$ is a uniform lattice in $\mathbb{R}$. As another example, consider the following explicit construction of a uniform lattice in $SL(2, \mathbb{R})$.

Example (2.1). Here we discuss two methods of explicit construction of a uniform lattice in $SL(2, \mathbb{R})$. First, we explain the geometric method. Suppose that X is a compact Riemann surface of genus greater or equal to two. For example, one may consider the Riemann surface of the function

$$\omega = \sqrt{z(z-1)(z-2)(z-3)(z-4)(z-5)}.$$

Let $\tilde{X}$ be the universal covering surface of X. It is known that $\tilde{X}$ is conformally equivalent to the upper half-plane $\mathbb{H}$, i.e, we may assume that $\tilde{X} \cong \mathbb{H}$. Then, by the theory of Riemann surfaces [106], we know that there is a discrete subgroup $\Gamma \subset PGL(2, \mathbb{R})$, such that

$$X \cong \Gamma \backslash \mathbb{H}.$$

Thus, Γ is a uniform lattice in $PGL(2, \mathbb{R})$, and one can now construct such a lattice in $SL(2, \mathbb{R})$. Recall that then Γ does not contain parabolic nor elliptic elements. The second method, is an algebraic method and uses elementary properties of indefinite divison quartenion algebras, (cf. [97]).

Let us now assume that G is any unimodular closed subgroup of $GL(r, \mathbb{R})$, i.e, G is a unimodular real Lie group. Let $\mathcal{H}_0$ be the Hilbert space

$$\mathcal{H}_0 = L^2(\Gamma \backslash G).$$

One can define a structure of a G-module on $\mathcal{H}_0$ by letting the unitary regular representation R to act on $\mathcal{H}_0$ as

$$(R(y)\varphi)(x) = \varphi(xy), \qquad (x, y \in G, \quad \varphi \in \mathcal{H}_0).$$

Let $f \in C_c^\infty(G)$, define

$$R(f) = \int_G f(x)R(x)dx.$$

Then $R(f)$ operates on $\mathcal{H}_0$ as follows:

$$\begin{aligned}(R(f)\varphi)(x) &= \int_G f(y)(R(y)\varphi)(x)dy, \qquad (\varphi \in \mathcal{H}_0)\\ &= \int_G f(y)\varphi(xy)dy\\ &= \int_G f(x^{-1}y)\varphi(y)dy,\end{aligned}$$

where the last equality is a consequence of the left invariance of dy. Using the Fubini theorem, the integral above can be written as

$$\begin{aligned}(R(f)\varphi)(x) &= \int_{\Gamma\backslash G} \left(\textstyle\sum_{\gamma\in\Gamma} f(x^{-1}\gamma y)\varphi(\gamma y)\right) dy\\ &= \int_{\Gamma\backslash G} \left(\textstyle\sum_{\gamma\in\Gamma} f(x^{-1}\gamma y)\right) \varphi(y)dy\\ &= \int_{\Gamma\backslash G} K(x, y)\varphi(y)dy,\end{aligned}$$

where

$$K(x,y)=\sum_{\gamma\in\Gamma} f(x^{-1}\gamma y).$$

This finite sum is called, *Selberg's kernel function.*

We now summarize some of the properties of $K(x,y)$ in the following lemma.

Lemma (2.1). The Selberg kernel function satisfies the following properties:

(i) $K(\gamma_1 x,\gamma_2 y)=K(x,y), \qquad (\gamma_1,\gamma_2\in\Gamma).$

(ii) $K(x,y)$ is a differentiable function on the compact manifold

$$(\Gamma\backslash G)\times(\Gamma\backslash G).$$

(iii) $R(f)$ is an integral operator whose kernel is $K(x,y)$.

Before stating the main proposition concerning the trace of an operator on a compact manifold, let us recall that $\mathcal{H}_0$ is an infinite dimensional vector space, and that, one defines the trace $\mathrm{tr}(R(f))$ as being the limit of the infinite series

$$\sum_{i=1}^{\infty} < R(f)\varphi_i,\varphi_i >,$$

where φ_i is an orthonormal basis of $\mathcal{H}_0$, and $<,>$ is the inner product on $\mathcal{H}_0$. If this series converges, one says that $R(f)$ is of *trace class* or it is a *trace class operator.* Moreover, by the known results of Functional Analysis we have the following lemma.

Lemma (2.2). Any trace class operator is a Hilbert-Schmidt operator, and any Hilbert-Schmidt operator, is a compact operator.

We now state the main proposition of the trace class operators on a compact manifold.

Proposition (2.3). Suppose that M is a compact manifold with a regular Borel measure $d\mu$. Let $h(x,y)$ be a smooth function on $M\times M$. Then, the operator

$$\varphi \xrightarrow{T} \int_M h(x,y)\varphi(y)d\mu(y),$$

where $\varphi\in L^2(M,d\mu)$ is of trace class, and its trace is given by

$$\mathrm{tr}(T)=\int_M h(x,x)d\mu(x).$$

In other words, the trace is given by the integral over the diagonal elements.

If we apply the proposition above to the operator $R(f)$, we can derive the Selberg trace formula, ie., we get

$$\mathrm{tr}\,R(f)=\int_{\Gamma\backslash G} K(x,x)dx=\int_{\Gamma\backslash G}\sum_{\gamma\in\Gamma} f(x^{-1}\gamma x)dx, \tag{2.1}$$

where, $f\in C_c^\infty(G)$. This formula has some other forms, which are obtained as follows.

Let $\{\gamma\}$ denote a representative set of the conjugacy classes in Γ, and $\Gamma_\gamma = \{\gamma_1 \in \Gamma : \gamma_1\gamma\gamma_1^{-1} = \gamma\}$ be the centralizer of γ in Γ. Then,

$$\operatorname{tr} R(f) = \int_{\Gamma\backslash G} \sum_{\{\gamma\}} \sum_{\delta\in\Gamma_\gamma\backslash\Gamma} f(x^{-1}\delta^{-1}\gamma\delta x)dx.$$

Use the finiteness of the inner sum to get

$$\operatorname{tr} R(f) = \sum_{\{\gamma\}} \int_{\Gamma\backslash G} \sum_{\delta\in\Gamma_\gamma\backslash\Gamma} f(x^{-1}\delta^{-1}\gamma\delta x)dx.$$

Then, replace the inner sum by an integral with counting measure $d\delta$, we get

$$\operatorname{tr} R(f) = \sum_{\{\gamma\}} \int_{\Gamma\backslash G} \int_{\Gamma_\gamma\backslash\Gamma} f(x^{-1}\delta^{-1}\gamma\delta x)d\delta dx.$$

Hence, by writing $t = \delta x$, we have

$$\operatorname{tr} R(f) = \sum_{\{\gamma\}} \int_{\Gamma_\gamma\backslash G} f(t^{-1}\gamma t)dt.$$

Now, if $G_\gamma = \{u \in G : \gamma u = u\gamma\}$ is the centralizer of γ in G, then the above equality can be written as

$$\operatorname{tr} R(f) = \sum_{\{\gamma\}} \int_{G_\gamma\backslash G} \int_{\Gamma_\gamma\backslash G_\gamma} f(t^{-1}u^{-1}\gamma ut)dudt,$$

where $u \in G_\gamma$. So, the final formula can be obtained by setting $x = ut$. Hence,

$$\operatorname{tr} R(f) = \sum_{\{\gamma\}} \operatorname{vol}(\Gamma_\gamma\backslash G_\gamma) \int_{G_\gamma\backslash G} f(x^{-1}\gamma x)dx, \tag{2.2}$$

where, dx is considered as the invariant measure on $G_\gamma\backslash G$. The existence of this invariant measure is explained below.

The equality (2.2) represents one of the forms of *Selberg's trace formula*, where the right-hand side summation will be called the *geometric expansion*. In fact, as one can see, the geometric expansion is a linear combination, (not necessarily finite), of the integrals of $f(x^{-1}\gamma x)$ over $G_\gamma\backslash G$. Such an integral is called an *orbital integral*. This name is justified by noting that $G_\gamma\backslash G$ is homeomorphic to the orbit $G\cdot\gamma = \{x^{-1}\gamma x : x \in G\}$ under the action of G on itself, by conjugation. It is known that, the isotropy group, (ie., the centralizer G_γ), is a unimodular group, and thus the quotient space $G_\gamma\backslash G$ carries a G-invariant measure. It is now clear that the main ingredient of the geometric expansion of Selberg's trace formula is the orbital integral

$$f_G(\gamma) = \int_{G_\gamma\backslash G} f(x^{-1}\gamma x)dx.$$

Here, we summarize some properties of the orbital integral $f_G(\gamma)$. To this end, we first give a definition.

Definition (2.4). Let G be a locally compact group, and $C(G)$ a space of functions on G. We say that D is a *distribution* on $C(G)$ if D is a linear functional on $C(G)$. Moreover, we say that D is an *invariant distribution*, if it is a distribution, and that

$$D(f^y) = D(f),$$

where

$$f^y(x) = f(yxy^{-1}), \qquad (x, y \in G).$$

Proposition (2.5). Let G be a closed subgroup of $GL(r, \mathbb{R})$. Then
(i) $f_G(\gamma)$ is a convergent integral.
(ii) $f_G(\gamma)$ is an invariant distribution on $C_c^\infty(G)$.

Proof. For a general proof of (i) see [87]. For (ii), note that

$$\begin{aligned}\int_{G_\gamma\backslash G} f^y(x^{-1}\gamma x)dx &= \int_{G_\gamma\backslash G} f(y(x^{-1}\gamma x)y^{-1})dx \\ &= \int_{G_\gamma\backslash G} f(x^{-1}\gamma x)dx,\end{aligned}$$

where the last equality is obtained by the invariance of the measure. ∎

Observe that in all discussions above, we have not directly employed the compactness condition of $\Gamma\backslash G$, except for the derivation of Selberg's trace formula (2.2), which is based on Proposition (2.3). We now again, consider this condition and derive another equivalent expression for the Selberg trace formula. This expression will be termed spectral expansion, which relates the trace formula to the multiplicity of an irreducible constituent of the regular representation of G.

Using the fact that Γ is a uniform lattice in G, we see that the spectral decomposition of R on $L^2(\Gamma\backslash G)$ consists only of the discrete spectrum. Thus, R is a direct sum of irreducible unitary representations. We consequently have

$$\begin{aligned}R &\cong \pi_1 \oplus \pi_1 \oplus \ldots \oplus \pi_2 \oplus \pi_2 \oplus \ldots \\ &\cong m_1\pi_1 \oplus m_2\pi_2 \oplus \ldots \\ &\cong \bigoplus_{\pi\in\hat{G}} m_\pi \pi,\end{aligned}$$

where $m_\pi \in \{0, 1, 2, \ldots\}$, and $\hat{G}$ is the *unitary dual* of G, i.e, the set of equivalence classes of irreducible unitary representations of G. We may write this isomorphism as

$$R(y) \cong \bigoplus_{\pi\in\hat{G}} m_\pi \pi(y), \qquad (y \in G).$$

Then, by multiplying both sides by $f(y) \in C_c^\infty(G)$, we get

$$R(f) \cong \bigoplus_{\pi\in\hat{G}} m_\pi \pi(f),$$

and

$$\operatorname{tr} R(f) = \sum_{\pi \in \hat{G}} m_\pi \operatorname{tr}(\pi(f)).$$

Hence, we can write another version of Selberg's trace formula as in (2.3), whose left-hand side expression will be called the *spectral expansion.*

$$\sum_{\pi \in \hat{G}} m_\pi \operatorname{tr}(\pi(f)) = \sum_{\{\gamma\}} \operatorname{vol}(\Gamma_\gamma \backslash G_\gamma) \int_{G_\gamma \backslash G} f(x^{-1}\gamma x)dx, \tag{2.3}$$

where as before, $f \in C_c^\infty(G)$.

Remark. Observe that we have not used any property of G being a real Lie group, except for assuming $f \in C_c^\infty(G)$. Thus, if we replace G by $G(\mathbf{A}), \Gamma$ by $G(\mathbb{Q})$, and suitably choosing a function f on $G(\mathbf{A})$, then the above Selberg trace formula holds for the pair $(G(\mathbf{A}), G(\mathbb{Q}))$. The choice of f will be explained in the next chapter.

2.3. Three examples

In his paper [93], Selberg mentions that his trace formula can be considered as a generalization of Poisson summation formula. This later formula, is an important tool in analytic number theory (cf. [109]) and in arithmetic (cf. [113]). Thus, it is important to give a proof of Poisson summation formula, using Selberg's trace formula, applied to the pair $(\mathbb{R}, \mathbb{Z})$. This will be explained in Example (2.6). Also, as it has been noted, Selberg's trace formula may be considered as a generalization of Frobenius reciprocity in finite group theory. We shall explain this in the second example of this section, applying Selberg's trace formula to the pair (G, Γ), where G is a finite group. The third example, which we shall not fully discuss, is due to Goldfeld [47], which shows how Weil's explicit formula [112] may be realized as Selberg trace formula for the pair $(\mathbf{A} \rtimes I_0, \mathbb{Q}^* \times \mathbb{Q})$, (cf. Example (2.8)).

Example (2.6). Suppose that $G = \mathbb{R}$, and $\Gamma = \mathbb{Z}$. Then, Γ is a discrete subgroup of $\mathbb{R}$, and that the quotient space $\mathbb{Z}\backslash\mathbb{R}$ is compact, being homeomorphic to a circle. Let $f \in C_c^\infty(\mathbb{R})$. Then we write down Selberg's trace formula (2.3), for $(\mathbb{R}, \mathbb{Z})$, and f. In fact, recall that $\hat{\mathbb{R}} = \{\pi_\lambda(y) = e^{i\lambda y} : \lambda \in \mathbb{R}\}$, then

$$\operatorname{tr}\pi_\lambda(f) = \int_{\mathbb{R}} f(y)e^{i\lambda y}dy.$$

This is, of course, the trace of the one-dimensional represenation $\pi_\lambda(f)$ which is the *Fourier transform* of f, i.e, we can write

$$\operatorname{tr} \pi_\lambda(f) = \hat{f}(\lambda).$$

Now, recall the multiplicity formula for π_λ (cf. Example (1.1) chapter I), we see that by Selberg's trace formula

$$\sum_{\pi \in \hat{\mathbb{R}}} m_\pi \operatorname{tr} \pi_\lambda(f) = \sum_{n \in \mathbb{Z}} \hat{f}(2\pi n) = \sum_{n \in \mathbb{Z}} f(n).$$

where $f \in C_c^\infty(\mathbb{R})$ and that the volume of $\mathbb{Z}\backslash\mathbb{R}$ is being normalized to 1. As it is clear, the formula above is the *Poisson summation formula.*

Example (2.7). Let G be a finite group and Γ any subgroup of G. Take any $\pi \in \hat{G}$ and let $f(x) = \operatorname{tr}\pi(x)$. The group G equipped with the discrete topology is a locally compact topological group such that $\Gamma\backslash G$ is compact, and one can take the counting measure as the invariant measure on G. Thus, observing that $\pi(f)$ is given exactly by the same formula as $R(f)$, we see in particular, that

$$\operatorname{tr}\pi(f) = \operatorname{tr}\left(\frac{1}{|G|}\sum_{x\in G} f(x)\pi(x)\right).$$

On the other hand, by Selberg's trace formula we can write

$$\sum_{\pi\in\hat{G}} m_\pi(\operatorname{tr}\pi(f)) = \frac{1}{|G|}\sum_{x\in\Gamma\backslash G}\sum_{\gamma\in\Gamma}\operatorname{tr}(\pi(x^{-1}\gamma x)).$$

By orthogonality of characters, the left-hand side of the equality above is m_π, and by a simple computation we can see that its right-hand side is

$$\frac{1}{|\Gamma|}\sum_{\gamma\in\Gamma}\operatorname{tr}\pi(\gamma),$$

so the equality can be written as

$$m_\pi = \frac{1}{|\Gamma|}\sum_{\gamma\in\Gamma}\operatorname{tr}\pi(\gamma).$$

Now, let $\pi|_\Gamma$ be the restriction of π to Γ, which is not necessarily an irreducible representation of Γ in $L^2(\Gamma\backslash G)$. Then, the left-hand side of the preceding formula is the multiplicity with which 1_Γ occurs in $\pi|_\Gamma$, and this is $[\pi|_\Gamma : 1_\Gamma]$. Moreover, the right hand side of the preceding equality is $[\pi : \operatorname{Ind}_\Gamma^G(1_\Gamma)]$. Hence, one obtains the equality

$$[\pi|_\Gamma : 1_\Gamma] = [\pi : \operatorname{Ind}_\Gamma^G(1_\Gamma)],$$

which is the *Frobenius reciprocity* in a special case.

In fact one should be able to modify the trace formula, and obtain the general form of the Frobenius reciprocity from the modified trace formula.

Example (2.8). Let $\mathbf{A}$ be the ring of adeles of $\mathbb{Q}$, and $\mathbf{A}^*$ the group of units of $\mathbf{A}$. Define the *ideles* of norm 1, to be the set

$$I_0 = \left\{x \in \mathbf{A}^* : |x| = \prod_p |x_p|_p = 1\right\}.$$

Then I_0 acts on $\mathbf{A}$ by left multiplication, and one can define the semi-direct product $\mathbf{A} \rtimes I_0$, whose multiplication operation is given by

$$(x, x')(y, y') = (xy, x'y + y').$$

Then the quotient space $\mathbb{Q}^* \times \mathbb{Q}\backslash\mathbf{A} \rtimes I_0$ is compact. Let $(x, x') \in \mathbf{A} \rtimes I_0$, and consider the space of functions f on $\mathbf{A} \rtimes I_0$, which are infinitely differentiable and of compact support with respect to x_∞ (i.e, the $\mathbb{R}$-component of x). Then, as it has been shown in [47] applying Selberg's trace formula to the pair $(\mathbf{A} \rtimes I_0, \mathbb{Q}^* \times \mathbb{Q})$ with respect to f, one deduces *Weil's explicit formula* [112]. (see also [110]).

2.4. A necessary condition

The purpose of this section is simply to show that the compactness condition on the quotient space $\Gamma\backslash G$ is necessary for the convergence of the geometric expansion of Selberg's trace formula. For this, we recall the integration formula on certain locally compact closed subgroup of $GL(2,\mathbb{R})$.

Suppose that

$$G = \{g \in GL(2,\mathbb{R}) : |\det g| = 1\} = GL(2,\mathbb{R})^1,$$

and

$$\Gamma = GL(2,\mathbb{Z}).$$

Let $G = NAK$, be the Iwasawa decomposition of G, where

$$N = \left\{\begin{pmatrix} 1 & u \\ 0 & 1 \end{pmatrix} : u \in \mathbb{R}\right\},$$

$$A = \left\{\begin{pmatrix} r & 0 \\ 0 & r^{-1} \end{pmatrix} : r > 0\right\},$$

$$K = O(2,\mathbb{R}), \quad \text{the orthogonal group.}$$

Then, for $f \in C_c^\infty(G)$,

$$\begin{aligned}\int_G f(x)dx &= \int_N \int_A \int_K f(nak)dndadk \\ &= \int_{-\infty}^{+\infty} \int_0^\infty \int_K r^{-2} f\left(\begin{pmatrix} 1 & u \\ 0 & 1 \end{pmatrix}\begin{pmatrix} r & 0 \\ 0 & r^{-1} \end{pmatrix} k\right) du\frac{dr}{r}dk.\end{aligned}$$

It is easy to verify that the right-hand side of the equality above is right-invariant by K, and left-invariant by N, and A. This is our integration formula on G. Now, observe that the right invariant measure on $N\backslash G$ is given by

$$\phi \to \int_0^\infty \int_K r^{-2}\phi\left(\begin{pmatrix} r & 0 \\ 0 & r^{-1} \end{pmatrix} k\right)\frac{dr}{r}dk,$$

where $\phi \in C_c^\infty(N\backslash G)$.

We now calculate the geometric expansion of Selberg's trace formula in two cases.

(i) Let $\gamma = \begin{pmatrix} 1 & 0 \\ 0 & -1 \end{pmatrix}$, then

$$G_\gamma = \left\{\begin{pmatrix} \pm r & 0 \\ 0 & \pm r^{-1} \end{pmatrix} : \ r > 0\right\},$$

and

$$\Gamma_\gamma = \left\{ \begin{pmatrix} \pm 1 & 0 \\ 0 & \pm 1 \end{pmatrix} \right\}.$$

Hence, $\Gamma_\gamma \backslash G_\gamma \cong \mathbb{R}^+$, and $\mathrm{vol}(\Gamma_\gamma \backslash G_\gamma) = +\infty$. This in particular means that, the geometric expansion of the Selberg trace formula is not defined. Observe that the choice of γ as above is due to the non-compactness of $\Gamma \backslash G$.

(ii) It is possible that the sum over $\{\gamma\}$ in the geometric expansion of Selberg's trace formula diverges. For example: let $\gamma_t = \begin{pmatrix} 1 & t \\ 0 & 1 \end{pmatrix}$ $t \in \mathbb{N}$, then, each γ_t forms a unique conjugacy class in Γ, and

$$G_{\gamma_t} = \left\{ \begin{pmatrix} 1 & u \\ 0 & 1 \end{pmatrix} : u \in \mathbb{R} \right\} \cup \left\{ \begin{pmatrix} -1 & u \\ 0 & -1 \end{pmatrix} : u \in \mathbb{R} \right\},$$

$$\Gamma_{\gamma_t} = \left\{ \begin{pmatrix} \pm 1 & k \\ 0 & \pm 1 \end{pmatrix} : k \in \mathbb{Z} \right\}.$$

Moreover,

$$\mathrm{vol}(\Gamma_{\gamma_t} \backslash G_{\gamma_t}) = \mathrm{vol}(\mathbb{Z} \backslash \mathbb{R}) = 1.$$

We now show that the geometric expansion of Selberg's trace formula with the above data diverges.

A simple computation shows that the geometric expansion is given by

$$\sum_t 1 \int_{G_{\gamma_t} \backslash G} f(x^{-1} \gamma_t x) dx.$$

On the other hand, N is a subgroup of finite index in G_{γ_t}, we then show that the sum

$$\sum_t 1 \int_{N \backslash G} f(x^{-1} \gamma_t x) d^* x, \qquad (t \in \mathbb{N}),$$

where d^*x is the invariant measure on $N \backslash G$, diverges. To this end, let

$$\begin{aligned} x^{-1} \gamma_t x &= k^{-1} a^{-1} n^{-1} \gamma_t n a k \\ &= k^{-1} a^{-1} \gamma_t a k. \end{aligned}$$

Then, the above summation can be written as

$$\sum_t \int_0^\infty \int_K r^{-2} f\left(k^{-1} \begin{pmatrix} r^{-1} & 0 \\ 0 & r \end{pmatrix} \begin{pmatrix} 1 & t \\ 0 & 1 \end{pmatrix} \begin{pmatrix} r & 0 \\ 0 & r^{-1} \end{pmatrix} k\right) \frac{dr}{r} dk =$$

$$= \sum_t \int_0^\infty \int_K f\left(k^{-1} \begin{pmatrix} 1 & r^{-2} t \\ 0 & 1 \end{pmatrix} k\right) r^{-3} dr dk.$$

Let $v = r^{-2} t$, then $dv = 2r^{-3} t\, dr$, or

$$r^{-3} dr = \frac{dv}{2t}.$$

Hence, the preceding summation can be written as

$$\sum_t \left(\int_K \int_0^\infty f \left(k^{-1} \begin{pmatrix} 1 & v \\ 0 & 1 \end{pmatrix} k \right) dvdk \right) \frac{1}{2t},$$

where $f \in C_c^\infty(G)$. Observe that, f has compact support, thus, the double integral inside the sum is finite. Therefore to get the divergence, one has to show that f may be chosen in such a way, that

$$\int_K \int_0^\infty f \left(k^{-1} \begin{pmatrix} 1 & v \\ 0 & 1 \end{pmatrix} k \right) dvdk \neq 0.$$

Indeed, this is of course possible. Now, the two cases (i) and (ii), explained above should be sufficient for the reader to see why Selberg's trace formula fails for a pair (G, Γ) when the quotient space $\Gamma \backslash G$ is not compact.

2.5 Generalizations and applications

An immediate question that arises after the study of the preceding section, is, how to develop a trace formula for more general connected groups, which bears Selberg's trace formula as an special case. This problem was studied by J. Arthur, and his works resulted in a trace formula that generalizes the one of Selberg. This will be explained in this book. However, there is another way to generalize Selberg's trace formula. We now explain it.

Suppose that G is a connected locally compact topological group, and Γ a uniform lattice in G. Let ϵ be an automorphism of G of finite order ℓ, and $< \epsilon >$ the group generated by ϵ. Then one can form the disconnected group

$$G^* = G \rtimes < \epsilon >,$$

defined by the semi-direct product of G and $< \epsilon >$. The elements of G^* are denoted by $g \times t$, where $g \in G$, $t \in < \epsilon >$, and the connected components of G^* are given by

$$G^t = G \times t.$$

In particular, if we assume that ϵ fixes Γ. i.e., $\epsilon(\Gamma) = \Gamma$, then $\Gamma \backslash G^\epsilon$ is compact. Now, define

$$(R^\epsilon(f)\varphi)(x) = \int_G f(y)\varphi(\epsilon^{-1}(xy))dy,$$

where f is a suitable function of compact support on G, and $\varphi \in L^2(\Gamma \backslash G)$. Then, we can calculate the trace of $R^\epsilon(f)$ on $L^2(\Gamma \backslash G)$, and the result is the *twisted Selberg trace formula* whose geometric expansion is based on the *twisted Selberg kernel function* $K^\epsilon(x, y)$

$$K^\epsilon(x, y) = \sum_{\gamma \in \Gamma} f(x^{-1}\gamma\epsilon(y)).$$

The gometric expansion of this trace formula is then a linear combination of the *twisted orbital integral*, defined by

$$\int_{G_{(\gamma\epsilon)} \backslash G} f(x^{-1}\gamma\epsilon(x))dx,$$

where the *twisted centralizer* $G_{(\gamma\epsilon)}$ is given by

$$G_{(\gamma\epsilon)} = \{x \in G : g = x^{-1} g \epsilon(x)\}.$$

A general twisted trace formula has been studied in [80] for an arbitrary reductive algebraic group over number fields. The idea was essentially to generalize the works of Arthur to the non-connected reductive algebraic groups (see also [60], [61]).

We now mention some known applications of Selberg's trace formula.

Several applications of Selberg's trace formula in analytic number theory are known, these applications are either a direct consequence of the trace formula, or, they are consequence of the Selberg zeta function (cf. [41], [52], [53], [95], [108], [109], [110]). In algebra its applications are directed to the study of quaternion algebras, and automorphic forms on the quaternion algebra. These series of applications in the modern treatments begin with the work of Jacquet and Langlands [57], (see also [42], [43]). Some of the applications of Selberg's trace formula in analysis can be found in the works of Faddeev, Lax, and Phillips [72]. In topology the applications of the trace formula is via representation theory. To see this, recall that there are topological meaning for multiplicity of an irreducible representation [75], [76], [77], [98]. On the other hand, as we have seen the multiplicity of an irreducible unitary representation of a locally compact group is an ingredient of the spectral expansion of Selberg's trace formula. So, when Selberg's trace formula is applicable, and one can get informations on these multiplicities, one can equally obtain informations on topological notions, such as Betti numbers, and the Euler-Poincaré characteristic. However, there is another approach that relates Selberg's trace formula to some topological notions. This goes back to the classical work of Eichler. The idea is to apply two different methods to calculate the trace of Heche operators, one is by means of a Lefschetz trace formula [37], and the other by Selberg's trace formula [36], [93]. Then, the comparison of these results yields analytical informations for topological notions. In another instance, Langlands in his paper [69], shows how a version of Lefschetz trace formula can be used together with Selberg's trace formula to prove some very deep results on the arithmetic of modular curves.

We now discuss some applications of Selberg's trace formula in index theory, and arithmetical geometry.

Suppose that V is a vector space and L an operator on V. Assume that $KerL$, and Coker $L = V/ImL$ are finite dimensional vector subspaces, then, in general, the *index* of L asks for the difference of the

$$\text{index } L = \dim KerL - \dim CokerL.$$

In certain cases, however, the index is given by the difference of two traces, and so, when Selberg's trace formula is applicable, one can calculate the index, using the trace formula. For a recent work on this subject see [23], and the references therein.

In arithmetical geometry, Selberg's trace formula is used to solve problems related to the arithmetic of some Shimura varieties [71]. Also, let us consider the role of Selberg's trace formula in arithmetic varieties. Recall that an arithmetic subgroup Δ is *neat* if $\gamma \in \Delta$, and γ^m is unipotent for some $m > 0$, then γ itself is unipotent [25]. This implies that Δ is torsion free. For example, in $GL(r, \mathbb{Z})$, any congruence subgroup $\Gamma(N)$ is torsion free if $N \geq 3$ (this is a theorem of Minkowski [78]). Let, D be a symmetric domain, and G the identity connected component of the group of holomorphic automorphism of D onto itself. Then, G is a semisimple Lie group of hermitian type with the center reduced to the identity [89]. Suppose that Γ is a neat arithmetic subgroup in G. Then, the action of Γ on D has no fixed point in D, and the space $\Gamma\backslash D$ turns into a complex manifold with no singularity inside the quotient space, except possibly for the singularities in the boundary. One can now study automorphic or modular forms on D with respect to Γ, or on a suitable compactification of $\Gamma\backslash D$. For example, knowing that the dimension of the corresponding space of modular forms is finite, one can express the dimension of the space of cusp forms, using Selberg's trace formula. This fact was actually observed by Selberg [94], see also ([45], [70]). On the other hand, this dimension can be calculated by means of geometrical tools, and the theorem of Riemann-Roch. This idea was first applied by Hirzebruch [54], when the quotient space $\Gamma\backslash D$ is compact. The result which is known as *Hirzebruch proportionality*, shows that Chern numbers of the compact arithmetic variety $\Gamma\backslash D$ are proportional to the Chern numbers of its compact dual. But, when the variety $\Gamma\backslash D$ is non compact, there are singularities at infinity (cusps, or boundary components of higher ranks). In this case, again, one can apply Selberg's trace formula to calculate the dimension of the space of cusp forms (as we have mentioned above). Moreover one can apply the general Riemann-Roch theorem for algebraic manifolds [55], and together with some geometric methods to calculate the dimension. Then, the comparison of these two results, leads to a conjecture of Satake, named: *the generalized Hirzebruch conjecture*, (cf. [84], [91]). We shall return to this subject elsewhere.

Chapter III

Kernel Functions And The Convergence Theorem

The main goal of this chapter is to prove a convergence theorem for certain kernel function which is basic for the generalization of Selberg's trace formula. For this we need some preparations.

Recall that as we have seen in chapter II, a reason for the divergence of the geometric expansion in the Selberg's trace formula comes from the existence of the upper triangular elements in $GL(2,\mathbb{Z})$, with identity on the diagonal, i.e., the *unipotent* elements. In a more general setting, when the quotient space $\Gamma\backslash G$ is non-compact Γ always contains unipotent elements. Thus, to generalize Selberg's trace formula one needs to define a suitable kernel function that generalizes the one of Selberg, such that its integral over $\Gamma\backslash G$ converges. Observe that, unipotent elements are in parabolic subgroups and, parabolic subgroups govern the geometry of $\Gamma\backslash G$ in the boundary [22]. Thus one does not have a single generalized kernel function, and in fact, there is a family of these kernels, which are parametrized by parabolic subgroups. This subject will be discussed in the second section. In the first section we review some basic facts about $GL(r)$ and its subgroups; in the last two sections we discuss some facts on reduction theory, and the convergence theorem, respectively.

3.1. Preliminaries on $GL(r)$

In the first chapter of this book we have already mentioned the relation between the partitions of a given positive integer r and $GL(r)$. In fact, let $(r_1,\dots r_\ell)$ be a partition of $r\in\mathbb{Z}$, $r>0$. To this partition one associates the group P_ℓ which consists of square matrices of size r of the form

$$\begin{pmatrix} \boxed{*} & & * \\ & \ddots & \\ 0 & & \boxed{*} \end{pmatrix} \begin{matrix} \}r_1 \\ \vdots \\ \}r_\ell \end{matrix}$$

The group P_ℓ will be called *the parabolic subgroup associated to the partition ℓ*. Inside P_ℓ, there is a subgroup which is called the *Levi component* of P_ℓ, and it is denoted by M_ℓ, which consists of matrices

$$\begin{pmatrix} \boxed{*} & & 0 \\ & \ddots & \\ 0 & & \boxed{*} \end{pmatrix} \begin{matrix} \}r_1 \\ \vdots \\ \}r_\ell \end{matrix} .$$

Let, M_ℓ^1 be the set of all elements of $m\in M_\ell$ such that

$$|\det m_i| = 1 \quad \forall i(i=1,\dots \ell),$$

where m_i is the ith block matrix of m. Then, M_ℓ^1 is a subgroup of M_ℓ. Now, let A_ℓ be the subgroup of the diagonal matrices

$$\begin{pmatrix} \begin{matrix} a_1 & & \\ & \ddots & \\ & & a_1 \end{matrix} & & 0 \\ & \ddots & \\ 0 & & \begin{matrix} a_\ell & & \\ & \ddots & \\ & & a_1 \end{matrix} \end{pmatrix} \begin{matrix} \}r_1 \\ \vdots \\ \}r_\ell \end{matrix}$$

with $a_i > 0$. Then, we see that

$$M_\ell = M_\ell^1 A_\ell.$$

The Group A_ℓ is called the *split component* of P_ℓ. Then, define the *unipotent radical* of P_ℓ to be the group of all matrices of the form

$$\begin{pmatrix} I_{r_1} & & * \\ & \ddots & \\ 0 & & I_{r_\ell} \end{pmatrix} \begin{matrix} \}r_1 \\ \vdots \\ \}r_\ell \end{matrix}$$

From this we see that

$$P_\ell = M_\ell N_\ell = M_\ell^1 A_\ell N_\ell.$$

Observe that it is an standard fact from linear algebra that, by using Gram-Schmidt orthogonalization, one can write any non-singular matrix as a product of an upper-triangular non-singular matrix, and an element in an orthogonal group K (see [88]). Thus, if P_0 is the subgroup of the upper-triangular matrices in

$$G = GL(r, \mathbb{R})^1 = \{g \in GL(r, \mathbb{R}) : |\det g| = 1\},$$

then

$$G = P_0 K.$$

Hence, for any P_ℓ as above, we have

$$G = P_\ell G = P_\ell P_0 K = P_\ell K,$$

since $P_0 \subset P_\ell$. Finally, we have the following decomposition for G

$$G = P_\ell K = M_\ell N_\ell K = M_\ell^1 A_\ell N_\ell K.$$

Therefore, any element $x \in GL(r, \mathbb{R})^1$ can be written as

$$x = pk_1 = mnk_2 = m^1 an_1 k_3,$$

where, $p \in P_\ell$, k_1, k_2, $k_3 \in K$, $m \in M_\ell$, $m^1 \in M_\ell^1$, and $a \in A_\ell$. Here, we mention that the following properties hold.

(i) P_ℓ is normalizer of N_ℓ, i.e.,

$$P_\ell = N_G(N_\ell) = \{g \in G : g^{-1} N_\ell g = N_\ell\}.$$

(ii) N_ℓ is normalized by A_ℓ, i.e.,

$$A_\ell \subset N_G(N_\ell).$$

(iii) P_ℓ is its own normalizer, i.e.,

$$P_\ell = N_G(P_\ell) = \{g \in G : g^{-1} P_\ell g = P_\ell\}.$$

(iv) M_ℓ is the centralizer of A_ℓ, i.e.,

$$M_\ell = C_G(A_\ell) = \{g \in G : gag^{-1} = a \quad \forall a \in A_\ell\}.$$

(v) P_ℓ, A_ℓ, M_ℓ, N_ℓ, M_ℓ^1, are all algebraic groups over $\mathbb{Q}$.

Now, say that a subgroup of G is *parabolic*, if it is conjugate in G to a subgroup P_ℓ.
Suppose that P is a parabolic subgroup of G. Denote by A_p (or A), N_P(or N), and M_p (or M), the split component, the unipotent radical, and the Levi component of P, respectively. These subgroups are conjugate to A_ℓ, N_ℓ, M_ℓ, respectively. Indeed, if $P = g^{-1}P_\ell g$, we set $A_P = g^{-1}A_\ell g$, $N_P = g^{-1}N_\ell g$, and $M_P = g^{-1}M_\ell g$. Then, we see that by $P_\ell = M_\ell N_\ell$, one has

$$\begin{aligned} g^{-1}P_\ell g &= g^{-1}M_\ell N_\ell g \\ &= g^{-1}M_\ell g g^{-1} N_\ell g. \end{aligned}$$

Thus,

$$P = MN.$$

Similarly

$$P = M_P^1 A_P N_p.$$

Moreover, if we write $G = g^{-1}Gg = g^{-1}P_\ell g^{-1} gKg$, then

$$G = PK_1,$$

where K_1 is a maximal compact subgroup of $GL(r, \mathbb{R})^1$ which is conjugate (by g) to K, and it is defined by

$$K_1 = \{k_1 \in G : {}^t k_1 (g\, {}^t g)^{-1} k_1 = g\, {}^t g\}.$$

We then by abuse of language write

$$G = PK,$$

where K is a maximal compact subgroup.

Example (3.1). The subgroup of the lower-triangular matrices of $GL(r, \mathbb{R})^1$ is a parabolic subgroup, being conjugate to the group of upper triangular matrices. See the diagram

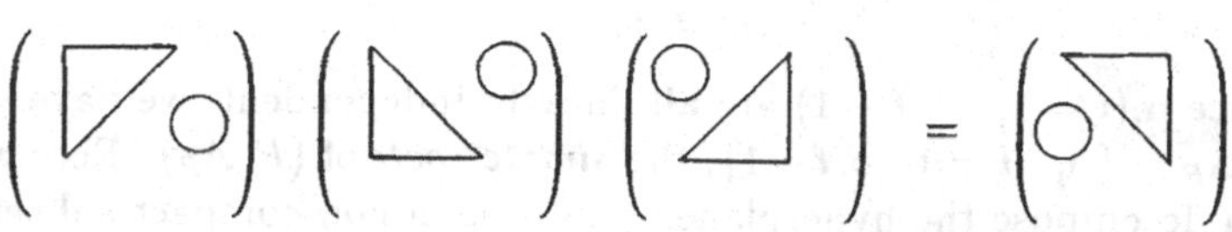

Now, suppose that a partition $(r_1, \ldots r_\ell)$ of r is fixed, and P is a parabolic subgroup associated to that partition. Let A_p, M_p be as above, then, denote by $X(M_P)$ the set of all characters of M_p (i.e., the homomorphisms of M_P onto $\mathbb{C}^*$ which are defined over $\mathbb{Q}$). Let,

$$m_P = \mathrm{Hom}(X(M_P), \mathbb{R}).$$

Then, m_P is a real vector space of finite dimension, whose dimension equals to that of A_P. Moreover, the dual vector space of m_P is denoted by m_P^*, whose dimension equals to that of m_P, and it is identified with

$$m_P^* = X(M_P) \otimes \mathbb{R}.$$

Similarly, one can define $X(A_P)$, and set

$$\mathbf{a}_P = \mathrm{Hom}(X(A_P), \mathbb{R}),$$
$$\mathbf{a}_P^* = X(A_P) \otimes \mathbb{R}.$$

We now define the important H_P-function. First, observe that the dimension of A_P is $\ell - 1$, equals to that of $\mathbf{a}_P$. Hence, in $\mathbb{R}^\ell$, $\mathbf{a}_P$ is an hyperplane, and it is identified with

$$\mathbf{a}_P = \{u = (u_1, \ldots u_\ell) \in \mathbb{R}^\ell : \sum_{i=1}^{\ell} u_i = 0\}.$$

Now, define the *H_P-function* by

$$H_P : G \to \mathbf{a}_P$$

given by

$$H_P(x) = (\log a_1, \ldots, \log a_\ell),$$

where $x \in G$, $x = m^1 an$, and

$$a = \mathrm{diag}(a_1, \ldots, a_\ell) \in A_P.$$

Observe that, since

$$\begin{aligned} \sum_{i=1}^{\ell} \log a_i &= \log \prod_{i=1}^{\ell} a_i \\ &= 0, \end{aligned}$$

H_p takes elements of G into $\mathbf{a}_p$. Moreover, since $|\det m_i| = a_i^{r_i}$, we have

$$H_P(x) = (\frac{1}{r_1}|\det m_1|, \ldots, \frac{1}{r_\ell}|\det m_\ell|).$$

Let us define the simple roots Δ_p of (P, A_P). First, define

$$\alpha_i : \mathbf{a}_P \to \mathbb{R} \qquad (i = 1, \ldots, \ell - 1),$$

by

$$\alpha_i(u) = u_i - u_{i-1}.$$

Note that, since $\alpha_i(i = 1,\ldots,\ell-1)$ are all linearly independent, we have shown that a basis of $\mathbf{a}_p$ is $\Delta_P = \{\alpha_i : i = 1,\ldots \ell-1\}$, the *simple roots* of (P, A_P). Fixing an ordering on Δ_p, we can decompose the hyperplane $\mathbf{a}_p$ into open non-compact subsets. Choosing the ordering of the positive numbers, we can define the *positive Weyl chamber* in $\mathbf{a}_p$ by

$$W_P^+ = \{u \in \mathbf{a}_P : \alpha_i(u) > 0 \quad \forall i = 1,\ldots,\ell-1\}.$$

Then, we say that

$$W_P^{i,+} = \{u \in \mathbf{a}_P : \alpha_i(u) = 0\}$$

is the *i-th wall* of the Weyl chamber. The *dual chamber* $\tilde{W}_p^+$ of W_p^+ is then defined as follows. First, for any $u \in \mathbb{R}^\ell$, define

$$\tilde{\omega}_i(u) = \sum_{t=1}^{i} n_t u_t - \sum_{t=i+1}^{\ell} n_t u_t.$$

Then, define

$$\tilde{W}_P^+ = \{u \in \mathbb{R}^\ell : \tilde{\omega}_i(u) > 0, \quad i = 1,\ldots,\ell-1\}.$$

To complete our notations from Lie theory begin by recalling that

$$\hat{\Delta}_P = \{\tilde{\omega}_i : i = 1,\ldots,\ell-1\}$$

is the set of *fundamental weights* of P. Now, suppose that there is a parabolic subgroup $Q \supset P$. Then, the partition of r to which Q is associated is $(r_1 + \ldots + r_{\nu_1}, r_{\nu_1+1} + \ldots + r_{\nu_2}, \ldots r_\ell)$. In other words, there is a one-to-one correspondence between the sets of parabolic subgroups Q containing P, and the partitions $(\ell_1,\ldots,\ell_q)$ of ℓ. Thus, $\mathbf{a}_Q$ is identified with

$$\mathbf{a}_Q = \{u = (u_{\ell_1}, u_{\ell_2}, \ldots u_{\ell_q}) \in \mathbb{R}^q : \sum_{i=1}^{q} u_{\ell_i} = 0\}.$$

It is clear that $\mathbf{a}_Q$ is a subspace of $\mathbf{a}_P$, and that there is a subspace $\mathbf{a}_P^Q$ such that

$$\mathbf{a}_P = \mathbf{a}_P^Q \oplus \mathbf{a}_Q.$$

Now, let

$$\Delta_P^Q = \{\alpha \in \Delta_p : \alpha|_{\mathbf{a}_Q} = 0\},$$

and set

$$\hat{\Delta}_P^Q = \text{ restriction of } \hat{\Delta}_P - \hat{\Delta}_Q \text{ to } \mathbf{a}_P^Q.$$

In particular, one denotes by τ_P (resp. $\hat{\tau}_P$) the *characteristic function* of W_P^+ (resp. $\tilde{W}_P^+$).

3.2. Combinatorics and reduction theory

The characteristic functions τ_P and $\tilde{\tau}_P$ just introduced, are the basic tools for the definition of the generalized kernel functions. The crucial role in the definition of these kernel functions is played by reduction theory of $\Gamma\backslash G$. Then, the proof of the convergence theorem for the generalized kernel function is a consequence of the reinterpretation of the reduction theory arguments in a combinatorial form related to these caracteristic functions. We now prove the following result for $\hat{\tau}_P$. For this purpose, we say that $T \in W_P^+$ is a *regular element* if it is sufficiently away from the walls of the positive Weyl chamber.

Lemma (3.2). Let $G_2 = GL(2,\mathbb{R})^1$, P a parabolic subgroup of G_2, $T \in W_P^+$ a regular element, and $\Gamma_2 = GL(2,\mathbb{Z})$. Then, for any $x \in G_2$, one has

$$\sum_{\delta \in \Gamma\cap\Gamma_2\backslash\Gamma_2} \hat{\tau}_P(H_P(\delta x) - T) < \infty.$$

Proof. We have only to discuss the partition $(1,1)$ of $r = 2$. Thus, there is only one proper parabolic subgroup, namely the minimal parabolic subgroup P_0. Let $T = (t_1, t_2)$ be an element in W_P^+, such that $t_1 - t_2 \gg 0$ (read sufficiently greater than 0), i.e., T is a regular element. For the group under consideration, we are able to show that the sum, consists only of one term, thus the finitness follows. For this, suppose that for an element δ, $\hat{\tau}_P(H(\delta x) - T) = 1$. Let, $y = \delta x$, then

$$y = n\begin{pmatrix} r & 0 \\ 0 & r^{-1} \end{pmatrix} k,$$

where $r > 0$, $k \in O(2,\mathbb{R})$ (the orthogonal group of 2×2 real matrices). Calculating $H_P(\delta x)$, we get

$$H_P(\delta x) = (\log r, -\log r).$$

Moreover,

$$H_P(\delta x) - T = (\log r - t_1, -\log r - t_2).$$

Then,

$$\begin{aligned} \tilde{\omega}(H_P(\delta x) - T) &= \log r - t_1 + \log r + t_2 \\ &= 2\log r - (t_1 - t_2). \end{aligned}$$

But, to have $H_P(\delta x) - T \in \tilde{W}_P^+$, we need

$$\tilde{w}(H_P(\delta x) - T) = 2\log r - (t_1 - t_2) > 0,$$

where $\tilde{w} = \tilde{w}_1$. Let now, δ_1 be another element in Γ_2 such that $\hat{\tau}_P(H_P(\delta_1 x) - T) = 1$; write $\delta_1 x = \eta y$, where $\eta = \delta_1\delta^{-1} \in \Gamma_2$. Calculating the Euclidean norm of the vector $y^{-1}\begin{pmatrix} 1 \\ 0 \end{pmatrix}$, we get

$$\|y^{-1}\begin{pmatrix} 1 \\ 0 \end{pmatrix}\| = \|k^{-1}\begin{pmatrix} r^{-1} & 0 \\ 0 & r \end{pmatrix}\begin{pmatrix} 1 & t \\ 0 & 1 \end{pmatrix}\begin{pmatrix} 1 \\ 0 \end{pmatrix}\| = \|\begin{pmatrix} r^{-1} \\ 0 \end{pmatrix}\| = r^{-1},$$

where $k^{-1} \in O(2,\mathbb{R})$, $n = \begin{pmatrix} 1 & t \\ 0 & 1 \end{pmatrix}$. Hence,

$$\tilde{w}(H_P(\delta x)) = -2\log\|y^{-1}\begin{pmatrix} 1 \\ 0 \end{pmatrix}\|.$$

Similarly,

$$\tilde{w}(H_P(\delta_1 x)) = \tilde{w}(H_P(\eta y)) = -2\log \|y^{-1}\eta^{-1}\begin{pmatrix}1\\0\end{pmatrix}\|.$$

Now, suppose that $\eta \notin P \cap \Gamma_2$ (i.e., η and δ represents different classes in $P \cap \Gamma_2 \backslash \Gamma_2$). Then,

$$\eta^{-1} = \begin{pmatrix} * & * \\ s & * \end{pmatrix}, \quad (s \in \mathbb{Z} - \{0\}).$$

On the other hand, as we just showed,

$$\tilde{w}(H_P(\eta y)) = -2\log \|y^{-1}\eta^{-1}\begin{pmatrix}1\\0\end{pmatrix}\|,$$

where the right hand side can be written as

$$\begin{aligned} -2\log \|k^{-1}\begin{pmatrix} r^{-1} & 0 \\ 0 & r\end{pmatrix}\begin{pmatrix} 1 & t \\ 0 & 1\end{pmatrix}\begin{pmatrix} * \\ s\end{pmatrix}\| \\ = -2\log \|k^{-1}\begin{pmatrix} r^{-1} & 0 \\ 0 & r\end{pmatrix}\begin{pmatrix} * \\ s\end{pmatrix}\| \\ = -2\log \|\begin{pmatrix} * \\ rs\end{pmatrix}\|. \end{aligned}$$

Hence, keeping in mind that $s \in \mathbb{Z} - \{0\}$, we have

$$\begin{aligned} \tilde{w}(H_P(\eta y)) &= -2\log \|\begin{pmatrix} * \\ rs\end{pmatrix}\| \\ &\leq -2\log r. \end{aligned}$$

Thus,

$$-2\log r \geq \tilde{w}(H_P(\eta y)) > t_1 - t_2 \gg 0,$$

and, this means that,

$$\log r < -\frac{1}{2}(t_1 - t_2) \ll 0.$$

But, this statement is a contradiction, since $r \in \mathbb{Z}^+$. Hence, $\eta \in P \cap \Gamma_2$ and this shows that the sum in the lemma has just one term. ∎

Remark (3.3). For a proof of the lemma above in the general case we refer the reader to the paper ([3], Lemma 5.1). However one can give a direct proof for the case of $GL(r, \mathbb{R})^1$ and $\gamma = GL(r, \mathbb{Z})$ based on the proof given above, but one has to observe that in this case, one needs to work with the representation of $GL(r, \mathbb{R})^1$ on $\wedge^i(\mathbb{R}^r)$ (the exterior product). Then, the vector $\begin{pmatrix}1\\0\end{pmatrix}$ will be replaced by the ***highest weight vector*** $e_1 \wedge e_2 \wedge \ldots \wedge e_i$, where e_j is the j-th element of the standard basis of $\mathbb{R}^r$ over $\mathbb{R}$.

A fundamental idea behind the proof of the convergence theorem of section 3 below, is a reduction theory of $\Gamma \backslash G$. This reduction theory was actually developed by Langlands

[68] for the study of Eisenstein series. Langlands reduction theory was slightly modified by Arthur in order to obtain the necessary language for the development of the trace formula. This is a reduction theory for the quotient space $P \cap \Gamma\backslash G$, where P is a parabolic subgroup of G. The essential idea is to divide the quotient space $P\cap\Gamma\backslash G$ into a compact subset and an unbounded subset, where the topology of the later is easy to understand. This is indeed a classical idea, as one can see through the *Siegel's reduction theory* for the pair $(\mathbb{H}, SL(2,\mathbb{Z}))$, or $(\mathbb{H}_r, Sp(2r,\mathbb{Z}))$, where $\mathbb{H}_r$ is the Siegel's upper half plane, and $Sp(2r,\mathbb{Z})$ is the integral symplectic matrices of size $2r$.

Let us review the necessary reduction theory needed for the development of the generalized trace formula. For this, first we recall the definition of the Siegel domain $S(T_0, C)$ in $GL(r, \mathbb{R})$.

Let P_0 be the minimal parabolic subgroup of $GL(r,\mathbb{R})$, i.e., the subgroup associated to the minimal partition $(1,1,\ldots,1)$ of r. Let, $N_0 = N_{P_0}$ be its unipotent radical, and C a compact subset of N_{P_0}. Then, in $\mathfrak{a}_0 = \mathfrak{a}_{P_0}$, fix a vector $T_0 = T_{P_0}$ which belongs to $\mathfrak{a}_0^+$. Now, define the *Siegel domain* $S(T_0, C)$ by:

$$\{nak \in CA_0K : \alpha(H_0(a) - T_0) > 0 \quad \forall \alpha \in \Delta_0\},$$

when, $A_0 = A_{P_0}$, $H_0(a) = H_{P_0}(a)$, $\Delta_0 = \Delta_{P_0}$.

Example (3.4). Let $G_3 = GL(3,\mathbb{R})$. Then P_0 is the subgroup of the upper-triangular matrices in G_3. N_0 is the subgroup of P_0 with diagonal elements 1. Let

$$C = \{n = (n_{ij}) \in N_0 : |n_{ij}| \leq \frac{1}{2} \quad (i < j)\}.$$

Suppose that $T_0 = (s, u, -s-u)$, where $s, u \in \mathbb{R}$. It is clear that $T_0 \in \mathfrak{a}_0$. Moreover, we have $T_0 \in \mathfrak{a}_0^+$. This means that $s > u$, and $u + s + u > 0$, i.e., $2u > -s$. Thus, $s > u > -\frac{s}{2}$. Now,

$$H_0(a) = (\log a_1, \log a_2, \log a_3),$$

where $a_i > 0$ $(i = 1,2,3)$. Then, we want

$$\alpha_1(H_0(a) - T_0) = \log a_1 - \log a_2 + u - s > 0,$$
$$\alpha_2(H_0(a) - T_0) = \log a_2 - \log a_3 - 2u - s > 0.$$

Hence,

$$\log\frac{a_1}{a_2} > s - u, \quad \log\frac{a_2}{a_3} > s + 2u.$$

Therefore, the diagonal matrix $a = \mathrm{diag}(a_1, a_2, a_3)$ of A_0 is determined by the inequalities:

$$\frac{a_1}{a_2} > \exp(s-u),$$
$$\frac{a_2}{a_3} > \exp(s+2u),$$
$$\frac{a_1}{a_3} > \exp(2s+u).$$

Hence, we have shown that $S(T_0, C)$ is an open *Siegel set* of G_3 (cf. [25]).

We now use $\tilde{w}$ to divide $S(T_0, C)$ into two parts, one relatively compact, and the other unbounded. For this, let $T \in \mathfrak{a}_0^+$ be a regular element. Then in $S(T_0, C)$ define the subset $S(T_0, T, C)$, by:

$$\{x \in S(T_0, C) : \tilde{w}(H_0(x) - T)) \leq 0 \ \ \forall \tilde{w} \in \tilde{\Delta}_0\}.$$

In particular, observe that this set is relatively compact in $GL(r, \mathbb{R})$, i.e., its closure is compact.

One is now in a position to define certain characteristic functions related to the domain $S(T_0, T, C)$. To begin, let $F(x, T)$ be the *characteristic function* of the projection of $S(T_0, T, C)$ onto $\Gamma\backslash G$. In the same way as above we can define $S^P(T_0, C)$, $S^P(T_0, T, C)$, for a parabolic subgroup $P \subset G$, but, we have to replace Δ_{P_0}, $\hat{\Delta}_{P_0}$ by $\Delta_{p_0}^P$, $\hat{\Delta}_{P_0}^P$ respectively. Then, for the Levi component $M = M_P$ of P, one defines the characteristic function of the image of $S^P(T_0, T, C)$ in $\Gamma \cap M\backslash M$. This function is denoted by $F^p(m, T)$. We now extend the function $F^P(m, T)$ to G. For an element $x \in G$, write $x = nmak$, where $n \in N_P$, $m \in M_P$, $a \in A_P$, and $k \in K$. Then, write

$$F^P(x, T) = F^P(m, T).$$

Of course in this way, one can regard $F^P(x, T)$ as a function on $\Gamma \cap P\backslash G$, since, it is left-invariant under $\Gamma \cap P$. A fundamental result about $F^P(x, T)$ is the following lemma whose proof will be postponed until next two chapters.

Lemma (3.5). For any $x \in G = GL(r, \mathbb{R})^1$, we have

$$\sum_P \sum_{\delta \in P\cap\Gamma\backslash\Gamma} F^P(\delta x, T)\tau_P(H_P(\delta x) - T)) = 1.$$

Let us look at this equality in more details. Observe that, to the trivial partition $\ell = 1$ of r i.e., $r = (r)$, it is associated the parabolic subgroup $P = G$. Thus, the summation in the left-hand side of the identity is given by

$$F^P(\delta x, T)\tau_G(H_G(\delta x) - T) + \sum_{P \neq G} \sum_{\delta \in P\cap\Gamma\backslash\Gamma} F^P(\delta x, T)\tau_P(H_P(\delta x) - T)).$$

The summation above represents the contribution from the boundary components. Moreover, we show that this is the only contribution. For this observe that if $P = G$, $\ell = 1$, $\mathfrak{a}_P = \{0\}$. Hence, $H_P(x) = 0$, and $T = 0$, (in fact, there is no regular element in $\mathfrak{a}_P$), and $W_P^+ = \phi$. Thus, $H_P(x) - T = 0$, and zero does not belong to W_P^+. This means that its characteristic function is zero, i.e.,

$$\tau_G(H_G(x) - T)) = 0.$$

Example (3.6). Since the function inside the summation is a characteristic function, this means that for a given x, there is exactly one P and one δ such that the function $F^P(x, T)\tau_P(H_P(\delta x) - T) = 1$.

As an example, consider $G_2 = GL(2, \mathbb{R})^1$, and $\Gamma_2 = GL(2, \mathbb{Z})$. Then, there are two parabolic subgroups P_0, and $P = G_2$ associated to the partitions (1,1), and (2) respectively. By what we have just seen, we can ignore the contribution from $P = G$, and if we write $P_0 \cap \Gamma_2 = \Gamma_\infty$, then the only contribution is

$$\sum_{\delta \in \Gamma_\infty \backslash \Gamma} F^{P_0}(\delta x, T)\tau_{P_0}(H_{P_0}(\delta x) - T).$$

But, $\Gamma_\infty \backslash \Gamma_2$ has only one element since, given $\gamma \in GL(2, \mathbb{Z})$, there is $\delta \in GL(2, \mathbb{Z})$ such that $\gamma\delta^{-1} \in \Gamma_\infty$. Thus, there is only one summand, namely

$$F^{P_0}(\delta x, T)\tau_{P_0}(H_{P_0}(\delta x) - T).$$

On the other hand, see that $H_0(\delta x) = H_0(a)$, where $a = \text{diag}(a_1, a_2)$, $a_1 > 0$, $a_2 = a_1^{-1}$. Also, observe that $\mathfrak{a}_0 = \{u = (u_1, u_2) \in \mathbb{R}^2 : u_1 + u_2 = 0\}$, and Δ_0 contains just one element α_1, given by $\alpha_1(u) = u_1 - u_2 = 2u_1$. The positive Weyl chamber is

$$\begin{aligned} W_0^+ &= \{u \in \mathfrak{a}_0 : \alpha_1(u) > 0\} \\ &= \{(u_1, -u_1) : u_1 > 0\}. \end{aligned}$$

Suppose that $T = (t, -t)$, with $t \in \mathbb{R}$, $t > 0$. Then,

$$H_0(\delta x) - T = (\log a_1 - t, -\log a_1 + t).$$

Hence, $\tau_0(H_0(\delta x) - T) = 1$, if $2\log a_1 - 2t > 0$, i.e., if $t < \log a_1$. Or equivalently

$$a_1 > e^t.$$

This is a restatement of the classical reduction theory [25].

Remark (3.7). Based on the example above, one can show that the identity of the lemma (3.5) is equivalent to the reduction theory of $\Gamma \backslash G$. For this one may begin by setting,

$$G_P(T) = \{x \in \Gamma \cap P \backslash G : F^P(x, T) = 1, \ \ \tau_P(H_P(x) - T) = 1\}.$$

Then,

(i) The projection

$$\varphi : G_P(T) \longrightarrow \overline{G_P(T)}$$

onto $\Gamma \backslash G$ is a homeomorphism.

(ii) $\Gamma \backslash G$ is the disjoint union over P of $\overline{G_P(T)}$.

Remark (3.8). One can investigate the relation between the reduction theory and Borel – Serre Compactification [29], (see also [117]), in the manner above.

We end up this section by describing some more combinatorial identities, and the methods which will be necessary for the proof of the convergence theorem in the next section. We begin by the simple, but very important identity given in the following exercise.

Exercise (3.9). Let $P \subset Q$ be parabolic subgroups. Show that if "card $\mathcal{U}$" denotes the cardinality of the finite subset $\mathcal{U}$, then

$$\sum_{\mathcal{U} \subset \Delta_P^Q} (-1)^{\text{card}\, \mathcal{U}} = \begin{cases} 0 & \text{if} \quad Q \neq P, \\ 1 & \text{if} \quad Q = P. \end{cases}$$

The next result from combinatorics is an identity about the characteristic functions τ and $\hat{\tau}$. More precisely, suppose that $P_1 \subset P_2$ are fixed parabolic subgroups, so that $\mathfrak{a}_{P_1} = \mathfrak{a}_{P_1}^{P_2} \oplus \mathfrak{a}_{P_2}$. Then, define

$$\sigma_{P_1}^{P_2}(H) = \sum_{Q \supset P_2} (-1)^{\dim(A_{P_2}/A_Q)} \tau_{P_1}^{Q}(H) \hat{\tau}_Q(H),$$

where, $H \in \mathfrak{a}_{P_1}$. From this we derive the following lemma (cf. Lemma 6.1, [3]).

Lemma (3.10). With the notations as above, $\sigma_{P_1}^{P_2}(H)$ is a characteristic function, i.e., its value is either 0 or 1. If $H = H_1^2 \oplus H_2$, where $H_1^2 \in \mathfrak{a}_{P_1}^{P_2}$, $H_2 \in \mathfrak{a}_{P_2}$, and $\sigma_{P_1}^{P_2}(H) = 1$, then $\tau_{P_1}^{P_2}(H_1^2) = 1$, and $\|H_2\| \leq c\|H_1^2\|$, where c is a constant. In particular, if $P_1 = P_2$, then $\sigma_{P_1}^{P_2}(H) = 0$.

3.3. The convergence theorem

The convergence theorem (3.17) that will be proved in this section, is for the integral of a kernel function that replaces the one of Selberg. To introduce this kernel function we first need to define an auxiliary kernel function parametrized by the parabolic subgroups of G. We do this as follows.

Let P be a parabolic subgroup of G. Write $P = MN$, then for a function $f \in C_c^\infty(G)$, and $x, y \in G$, define the *auxiliary kernel function*

$$K_P(x,y) = \int_N \sum_{\gamma \in M \cap \Gamma} f(x^{-1}\gamma n y)\, dn,$$

where, $n \in N$. Then, define the *Arthur kernel function*

$$k^T(x,f) = \sum_P (-1)^{\dim A} \sum_{\sigma \in P \cap \Gamma \backslash \Gamma} K_P(\delta x, \delta x) \hat{\tau}_P(H_P(x) - T_P), \tag{3.1}$$

where A is the split component of P, and

$$T_P = (t_1 + \ldots + t_{r_1}, t_{r_1+1} + \ldots + t_{r_2}, \ldots, \ldots + t_{r_\ell}) \in \mathfrak{a}_P.$$

Lemma (3.11). For the above kernel functions, the following properties hold.

(i) $K_P(x,y)$ is the kernel function of $(R_P(y)\phi_P)(x) = \phi_P(xy)$ on $L^2((M \cap \Gamma)N \backslash G)$.

(ii) $K_P(x,y) = \sum_{\gamma \in M \cap \Gamma} \int_N f(x^{-1}\gamma n y)\, dn.$

(iii) For $P = G$,

$$k^T(x,f) = K(x,x),$$

where the right hand side is Selberg's kernel function.

Proof. (i), one defines

$$(R_P(f)\phi_P)(x) = \int_G f(y)\phi_P(xy)\, dy,$$

where $f \in C_c^\infty(G)$, $x, y, \in G$. Then,

$$(R_P(f)\phi_P)(x) = \int_G f(x^{-1}y)\phi_P(y)dy,$$

where, the right-hand side integral can be written as

$$\int_{(M\cap\Gamma)N\backslash G}\int_{(M\cap\Gamma)N} f(x^{-1}ty)\phi_P(ty)dy\, dt.$$

Observe that, $t \in (M\cap\Gamma)N$, and dy is the induced Haar measure on $(M\cap\Gamma)N$. Then, $dt = d\gamma dn$, where, $d\gamma$ is the Haar measure on the discrete group $M\cap\Gamma$ (i.e., $d\gamma$ is the counting measure), and dn is the Haar measure on N, moreover, these measures are all normalized in a way that, Weil's integration formula on a group relative to a subgroup holds [114]. Then, the double integral above is written as

$$\int_{(M\cap\Gamma)N\backslash G}\int_N \sum_{\gamma\in M\cap\Gamma} f(x^{-1}\gamma ny)\phi_P(\gamma ny)dn\, dy$$
$$= \int_{(M\cap\Gamma)N\backslash G}\left(\int_N \sum_{\gamma\in M\cap\Gamma} f(x^{-1}\gamma ny)dn\right)\phi_P(y)dy.$$

This proves (i).

To prove (ii), observe that f has compact support, and $M\cap\Gamma$ is discrete. Thus the sum inside the integral is finite. This proves (ii).

The proof of (iii) is easy, since when $P = G$, then $N = \{I\}$, $M = G$, $M\cap\Gamma = \Gamma$. Thus,

$$\int_N f(x^{-1}\gamma nx)dn = f(x^{-1}\gamma x).$$

This implies that

$$K_G(x,x) = \sum_{\gamma\in\Gamma} f(x^{-1}\gamma x) = K(x,x).$$

∎

Lemma (3.12). Let P be a parabolic subgroup of G, and $T \in W_P^+$. Suppose that $C(T) \subset \Gamma\backslash G$ is a compact subset (of course, C depends on T). Then,

$$k^T(x,f) = K(x,x) \quad \forall x \in C(T).$$

Proof. One needs to show that for $P \not\subseteq G$, the terms of $k^T(x,f)$ corresponding to P vanish whenever $x \in C(T)$. ∎

We now go back to the kernel function $k^T(x,f)$. Our aim is to prove that

$$\int_{\Gamma\backslash G} |k^T(x,f)|dx < \infty.$$

The proof in the general case for reductive groups over number fields can be found in [3].

Now, we return to the definition of $k^T(x,f)$ as given in (3.1). For the coefficient "one" of $K_P(\delta x, \delta x)$ in the formula, substitute

$$\sum_{\{P_1:P_1\subset P\}} \sum_{\xi\in P_1\cap\Gamma\backslash P\cap\Gamma} F^{P_1}(\xi y, T)\tau_{P_1}^P(H_{P_1}(\xi y)-T)=1.$$

We obtain:

$$\begin{aligned} k^T(x,f) \;=\; & \sum_{\{P_1:P_1\subset P\}} (-1)^{\dim A_P} \sum_{\delta\in P_1\cap\Gamma\backslash\Gamma} F^{P_1}(\delta x,T)K_P(\delta x,\delta x)\times \\ & \tau_{P_1}^P(H_{P_1}(\delta x)-T)\hat{\tau}_P(H_P(\delta x)-T). \end{aligned} \tag{3.2}$$

Then by exercise (3.9) we have

$$\tau_{P_1}^P(x)\hat{\tau}_P(x)=\sum_{\{P_2,Q:P\subset P_2\subset Q\}} (-1)^{\dim(A_{P_2}/A_Q)}\tau_{P_1}^Q(x)\hat{\tau}_Q(x).$$

Since for fixed $Q\supset P$,

$$\sum_{\{P_2:P\subset P_2\subset Q\}} (-1)^{\dim(A_{P_2}/A_Q)} = \sum_{u\subset\Delta_P^Q} (-1)^{\operatorname{card} u},$$

the product $\tau_{P_1}^P(\cdot)\hat{\tau}_P(\cdot)$ in the identity (3.2) above represents the sum $\sum_{\{P_2:P_2\supset P\}} \sigma_{P_1}^{P_2}(x)$.

Hence, we have the following formula:

$$\begin{aligned} k^T(x,f) = & \sum_{\{P_1,P,P_2:P_1,\subset P\subset P_2\}} \sum_{\delta\in P_1\cap\Gamma\backslash\Gamma} \\ & (-1)^{\dim A_P} F^{P_1}(\delta x,T)\sigma_{P_1}^{P_2}(H_{P_1}(\delta x)-T)K_P(\delta x,\delta x). \end{aligned}$$

From which, by decomposing the first sum into two sums, we get:

$$\begin{aligned} \int_{\Gamma\backslash G} |k^T(x,f)|dx \;\le\; & \sum_{P_1\subset P_2} \int_{\Gamma\backslash G} \sum_{\delta\in P_1\cap\Gamma\backslash\Gamma} \Big|\sum_P (-1)^{\dim A_P} F^{P_1}(\delta x,T)\times \\ & \sigma_{P_1}^{P_2}(H_{P_1}(\delta x)-T)K_P(\delta x,\delta x)\Big| dx \\ \le\; & \sum_{P_1\subset P_2} \int_{P_1\cap\Gamma\backslash G} |A(x)||B(x)|dx, \end{aligned}$$

where

$$A(x)=F^{P_1}(x,T)\sigma_{P_1}^{P_2}(H_{P_1}(x)-T),$$

and

$$\begin{aligned} B(x) \;=\; & \sum_{\{P:P_1\subset P\subset P_2\}} (-1)^{\dim A_P} K_P(x,x) \\ =\; & \sum_P (-1)^{\dim A_P} \sum_{\gamma\in\Gamma\cap M_P} \int_{N_P} f(x^{-1}\gamma n\, x)dn. \end{aligned}$$

Remarks (3.13) (1) In the definition of $B(x)$, we can take the sum to be running over $\gamma\in\Gamma\cap M_P\cap P_1$ instead of over $\gamma\in\Gamma\cap M_P$.

(2) The proof of our theorem is based on the fact that both $A(x)$ and $B(x)$ are non-zero for some x. Indeed we shall remove all those x such that $B(x) = 0$, from $B(x)$. In fact such elements can be explicitly determined.

Let us illustrate this for the group $GL(2)$.

Consider the element $x = \begin{pmatrix} 1 & \alpha \\ 0 & 1 \end{pmatrix}\begin{pmatrix} r & 0 \\ 0 & r^{-1} \end{pmatrix} k$, where $r > e^{t_1-t_2} >> 0$. For $r = 2$, $P = P_0$, $P_1 = G$, $P_1 \neq P$, then, $\sigma_{P_1}^{P_2}(H_{P_1}(x) - T) = \tau_{P_0}(H_P(x) - T)$. Suppose $\gamma \in (\Gamma \cap M_P) - (\Gamma \cap M_P \cap P_1)$, where $\gamma = \begin{pmatrix} a & b \\ c & d \end{pmatrix}$ with $c \in \mathbb{Z} - \{0\}$. We have

$$f(x^{-1}\gamma n x) = f\left(k^{-1}\begin{pmatrix} * & * \\ r^2 c & * \end{pmatrix} k\right).$$

Since f has compact support and r^2c is very large ($r > e^{t_1-t_2} >> 0$), $f(x^{-1}\gamma n\, x)$ must be zero, otherwise if C denotes the support of f (which is compact) the inclusion

$$k^{-1}\begin{pmatrix} * & * \\ r^2 c & * \end{pmatrix} k \subset C$$

for large r^2c cannot hold. Therefore we can discard all such γ (i.e., we can take the sum over $\Gamma \cap M_P \cap P_1$). For an arbitrary r, the same idea is applicable (see [3] page 943).

Let $N_{P_1}^P = N_{P_1} \cap M_P = \exp(\mathcal{M}_{P_1}^P)$, where $\mathcal{M}_{P_1}^P \cong \mathbb{R}^{\dim N_{P_1}^P}$ is the Lie algebra of $N_{P_1}^P$. Let $\Lambda_{P_1}^P = \mathbb{Z}^{\dim N_{P_1}^P}$, the lattice in $\mathcal{M}_{P_1}^P$ corresponding to Γ, then

$$\begin{aligned}(M_P \cap \Gamma \cap P_1)N_P &= (M_{P_1} \cap \Gamma)(N_{P_1}^P \cap \Gamma)N_P \\ &= (M_{P_1} \cap \Gamma)\exp(\Lambda_{P_1}^P + \mathcal{M}_P).\end{aligned}$$

Therefore $B(x)$ can be written as

$$B(x) = \sum_{\gamma_1 \in M_{P_1} \cap \Gamma} \sum_{\{P: P_1 \subset P \subset P_2\}} (-1)^{\dim A_P} \sum_{\xi \in \Lambda_{P_1}^P} \int_{\mathcal{M}_P} f(x^{-1}\gamma_1 \exp(\xi + X)x)dX,$$

where $\gamma \cdot n = \gamma_1 \exp(\xi + X)$, $\mathcal{M}_{P_1} = \mathcal{M}_{P_1}^P \oplus \mathcal{M}_P$, $X \in \mathcal{M}_P$. Now apply the Poisson summation formula to the sum over $\xi \in \Lambda_{P_1}^P$ i.e., first, consider $\int_{\mathcal{M}_P} f(x^{-1}\gamma_1\exp(\xi + X)x)dX$, second look at its Fourier transform

$$\int_{\mathcal{M}_P} f(x^{-1}\gamma_1 \exp(X_1 + X)x)e^{2\pi i \xi\cdot(X_1+X)}dX_1,$$

third, take an integral of this with respect to $\mathcal{M}_{P_1}^P$, to get

$$\int_{\mathcal{M}_{P_1}^P}\int_{\mathcal{M}_P} f(x^{-1}\gamma_1 \exp(X_1 + X)x)e^{2\pi i \xi\cdot(X_1+X)}dX_1 dX.$$

Now sum over $\Lambda_{P_1}^{P_2}$ for ξ, and put $X_1 + X = Y$, to get

$$B(x) = \sum_{\gamma_1}\sum_{P}(-1)^{\dim A_P} \sum_{\xi \in \Lambda_{P_1}^P} \int_{\mathcal{M}_{P_1}} f(x^{-1}\gamma_1 \exp(Y)x)e^{2\pi i \xi\cdot Y}dY.$$

This implies that

$$|B(x)| \leq \sum_{\gamma_1 \in M_{P_1} \cap \Gamma} |\sum_{P} (-1)^{\dim A_p} \sum_{\xi \in \Lambda^P_{P_1}} \int_{M_{P_1}} f(x^{-1}\gamma_1 \exp(Y)x) e^{2\pi i \xi \cdot Y} dY|.$$

Next let P' be a minimal parabolic subgroup, with $P' \subset P$, $\Lambda^{P'}_{P_1} \subset \Lambda^P_{P_1} \subset \Lambda^{P_2}_{P_1}$, let $\xi \in \Lambda^{P'}_{P_1}$. The contribution of ξ to $|B(x)|$ will be

$$\sum_{\{P: P' \subset P \subset P_2\}} (-1)^{\dim A_p} = \begin{cases} 0 & \text{if } P' \subsetneq P_2 \\ 1 & \text{if } P' = P_2. \end{cases}$$

If we let $(\Lambda^{P_2}_{P_1})'$ be the set of elements in $\Lambda^{P_2}_{P_1}$ which do not belong to $\Lambda^P_{P_1}$ for any $P_1 \subseteq P \subsetneq P_2$, then we have shown that

$$|B(x)| \leq \sum_{\gamma \in M_{P_1} \cap \Gamma} \sum_{\xi \in (\Lambda^{P_2}_{P_1})'} |\int_{M_{P_1}} f(x^{-1}\gamma \exp(Y)x) e^{2\pi i \xi . Y} dY|.$$

From this we derive

$$\int_{\Gamma \backslash G} |k^T(x,f)| dx \leq \sum_{P_1 \subset P_2} \int_{P_1 \cap \Gamma \backslash G} dx \cdot (F^{P_1}(x,T) \sigma^{P_2}_{P_1}(H_{P_1}(x) - T)) \times$$

$$\sum_{\gamma \in M_P \cap \Gamma} \sum_{\xi \in (\Lambda^{P_2}_{P_1})} |\int_{M_{P_1}} f(x^{-1}\gamma \exp(Y)x) e^{2\pi i \xi \cdot Y} dY|.$$

Put

$$x = n^2_1 n_2\, mak, \tag{3.3}$$

where, $n^2_1 \in N^{P_2}_{P_1} = N_{P_1} \cap M_{P_2}$, $n_2 \in N_{P_2}$, $m \in M^1_{P_1}$, $a \in A_{P_1}$, $k \in K$. Then one can show that

$$dx = \delta_{P_1}(a) dn^2_1 dn_2 dm da dk,$$

where δ_{P_1} is the modular function of P_1. Moreover, $\delta_{P_1}(a)^{-1} dn = d(ana^{-1})$ with $n \in N_{P_1}$, $a \in A_{P_1}$, and we have $P_1 \cap \Gamma = (N^{P_2}_{P_1} \cap \Gamma)(N_{P_2} \cap \Gamma)(M^1_{P_1} \cap \Gamma)$. If we take n^2_1, n_2, m in a fixed fundamental domain in $N^{P_2}_{P_1}$, N_{P_2} and $M^1_{P_1}$ respectively, then

$$F^{P_1}(x,T)\sigma^{P_2}_{P_1}(H_{P_1}(x) - T) = F^{P_1}(m,T)\sigma^{P_2}_{P_1}(H_{P_1}(a) - T),$$

since the function $F^{P_1}(m,T)$ has compact support on $(M^1_{P_1} \cap \Gamma)\backslash M^1_{P_1}$. Also for the same reason n^2_1, n_2, m, k are integrated over compact sets. Our computation will then be continued by an awareness that the integral over n_2 is absorbed in the integral over Y and disappears, and we need only to consider points where the integrand does not vanish. Recall that if $\sigma^{P_2}_{P_1}(H_{P_1}(a) - T) \neq 0$, then $\tau^{P_2}_{P_1}(H_{P_1}(a) - T) \neq 0$. This implies that $a^{-1} n^2_1 a$ remains in a compact subset of $N^{P_2}_{P_1}$. Now we let x be as in (3.3). Then

$$f(x^{-1}\gamma \exp(Y)x) = f(k^{-1}a^{-1}m^{-1}(n^2_1)^{-1}\gamma \exp(Y) n^2_1 mak). \tag{3.4}$$

Moreover, if we put $y = (a^{-1}n^2_1 a)mk$, then (3.4) can be written as

$$f(y^{-1}a^{-1}\gamma \exp(Y)ay).$$

For this calculations see also page 944 of [3].

There is a compact subset Δ of G for which we have

$$\int_{\Gamma\backslash G} |k^T(x,f)|dx \leq \text{Const.} \sum_{P_1\subset P_2} \sum_{\gamma\in\Gamma\cap M_{P_1}} \sup_{y\in\Delta} \quad *$$

where $*$ equals the expression

$$\int_{A_{P_1}} \delta_{P_1}(a)^{-1}\sigma_{P_1}^{P_2}(H_{P_1}(a)-T) \sum_{\xi\in(\Lambda_{P_1}^{P_2})} |\int_{\mathcal{M}_{P_1}} f(y^{-1}a^{-1}\gamma\exp(Y)ay)e^{2\pi i\xi\cdot Y}dY|da.$$

But,

$$a^{-1}\gamma\exp(Y)a = \gamma a^{-1}\exp(Y)a = \gamma\exp(a^{-1}Ya),$$

and

$$e^{2\pi i\xi\cdot Y} = e^{2\pi i(a\xi a^{-1})\cdot(a^{-1}Ya)}.$$

Exercise (3.14). Check that the dot product has the property above.

Also, we have $\delta_{P_1}(a)^{-1}dY = d(a^{-1}Ya)$ and this implies

$$\sup_{y\in\Delta}\int_{A_{P_1}} \sigma_{P_1}^{P_2}(H_{P_1}(a)-T) \sum_{\xi\in(\Lambda_{P_1}^{P_2})} |\int_{\mathcal{M}_{P_1}} f(y^{-1}\gamma\exp(Y)y)e^{2\pi i(a\xi a^{-1})\cdot Y}dY|da$$

$$= \text{Const.} \sum_{P_1\subset P_2}\sum_{\gamma}\sup_{y}\int_{A_{P_1}} \sigma_{P_1}^{P_2}(H_{P_1}(a)-T) \sum_{\xi\in(\Lambda_{P_1}^{P_2})'} |F_{\gamma,y}(a\xi a^{-1})|da,$$

where,

$$F_{m,y}(X) = \int_{\mathcal{M}_{P_1}} f(y^{-1}m\exp(Y)y)e^{2\pi i(X\cdot Y)}dY,$$

and $m\in M_{P_1}$, $X\in\mathcal{M}_{P_1}$. In fact $F_{m,y}$ is a smooth function in y, compactly supported in m_1 as well as being a Schwartz function in X. We likewise require the following result of [3] page 939.

Corollary (3.15). Fix $T\in\mathfrak{a}_0^+ = \{H\in\mathfrak{a}_0 : \alpha(H)>0,\ \alpha\in\Delta_0\}$. For any $H\in\mathfrak{a}_{P_1}$, let H_1^2 be the projection of H onto $\mathfrak{a}_{P_1}^{P_2}$. Then if $\sigma_1^2(H-T)\neq 0$, $\alpha(H_1^2)>0\ \ \forall\ \alpha\in\Delta_{P_1}^{P_2}$, and $\|H\|\leq c(1+\|H_1^2\|)$, for any Euclidean norm $\|\cdot\|$ on $\mathfrak{a}_0$ and some constant c.

Then the projection of $H_{M_1}(a)-T$ onto $\mathfrak{a}_{P_1}$ can be written as

$$\begin{aligned} H_{M_1}(a)-T &= t_1\tilde{\omega}_1^v+\ldots+t_p\tilde{\omega}_p^v+H_2 \\ &= \textstyle\sum_{\alpha\in\Delta_{P_1}^{P_2}} t_\alpha\tilde{\omega}_\alpha^v+H_2, \end{aligned}$$

where, $t_i\in\mathbb{R}$, $H_2\in\mathfrak{a}_{P_2}$, and for each $\alpha\in\Delta_{P_1}^{P_2}$, $t_\alpha>0$. Now, if $\sigma_{P_1}^{P_2}(H_{M_1}(a)-T)\neq 0$, it follows from the corollary above that H_2 belongs to a compact set whose volume is limited by a polynomial in the numbers $\{t_\alpha\}$. More precisely, $\|H_2\|$ is bounded by a power of $|t_1|+|t_2|+\ldots+|t_p|$. Here our aim is to determine the last integral with respect to a and this amounts to the determination of the integral with respect to $dt_1\ldots dt_p\,dH_2$.

To do this, let $\xi \in \Lambda_{P_1}^{P_2}$, then $\xi \in \Lambda_{P_1}^{P_2}$ if and only if $\xi = \sum_{\beta} c_\beta X_\beta$, where β is a root of (G, A_{p_1}), $c_\beta \in \mathbb{Z}$, $X_\beta \in \mathcal{M}_\beta$, $\|X_\beta\| = 1$ and

$$\begin{aligned} aX_\beta a^{-1} &= e^{\beta(H_{M_1}(a))} X_\beta \\ &= C_T \exp(m_{\beta,1}t_1 + \ldots + m_{\beta,p}t_p) X_\beta. \end{aligned}$$

That is to say, $\beta = \sum_{i=1}^{p} m_{\beta,i}\alpha_i$, where $m_{\beta,i} \in \mathbb{N} - \{0\}$. But, $\xi \in (\Lambda_{P_1}^{P_2})'$ if and only if for any i with $1 \leq i \leq p$, there is β with $m_{\beta,i} \neq 0$ and $c_\beta \neq 0$. Then the convergence of the integral follows from the exercise below.

Exercise (3.16). Prove that if ϕ is a Schwartz function on $\mathbb{R}^n$, then there is a constant $c(\phi)$ such that for every lattice $\Lambda \subset \mathbb{R}^n$, with $\mathrm{vol}(\Lambda \backslash \mathbb{R}^n) \geq 1$, and for any N, $\sum_{\{\xi \in \Lambda : \|\xi\| \geq N\}} |\phi(\xi)| \leq \frac{c(\phi)}{N}$.

We have thus proved.

Theorem (3.17). With the notations as above, one has

$$\int_{\Gamma \backslash G} |k^T(x, f)| dx < \infty.$$

Chapter IV

The Adèlic Theory

The aim of this chapter, which consists of only one section, is to prepare the necessary back ground for the use of the adèlic language in the trace formula. In this setting, one can apply the trace formula in certain number theoretical questions [20], [58], [67].

4.1. Basic facts

We have already defined the notion of the adèle in the first chapter of this book. Here, we define the adèles, based on the notion of direct limit [38]. To be precise, let S be a finite subset of valuations $v \in \mathbb{Q}$ containing the Archimedean valuation $|\cdot|_\infty$. Let $\mathbb{Z}_v$ be the ring of integers of v, then set

$$\mathbf{A}_S = \{\prod_v a_v : a_v \in \mathbb{Q}_v, \;\; a_v \in \mathbb{Z}_v \text{ if } v \notin S\}.$$

Since $\mathbb{Z}_v$ in v-adic topology is compact, $\mathbf{A}_S$ is a locally compact ring. Now define the topological direct limit $\mathbf{A}$ of $\mathbf{A}_S$, i.e.,

$$\mathbf{A} = \varinjlim_S \mathbf{A}_S.$$

$\mathbf{A}$ is called the ring of *adèles* of $\mathbb{Q}$, it is a locally compact ring, and $\mathbb{Q} \subset \mathbf{A}$ is a discrete subring embedded in $\mathbf{A}$ by diagonal embedding. The adèlic groups can be also defined similarly. For example, if $G = GL(r)$, we may define $G(\mathbf{A})$ directly in the following way.

Identify the valuations of $\mathbb{Q}$ with the primes of $\mathbb{Q}$ (finite, and the infinite one). Then, for any finite prime $p \notin S$, define $K_p = G(\mathbb{Z}_p)$ (this is a compact group), and set

$$G(\mathbf{A}_S) = \prod_{p\in S} G(\mathbb{Q}_p) \prod_{p \notin S} K_p,$$

then

$$G(\mathbf{A}) = \varinjlim_S G(\mathbf{A}_S).$$

In $G(\mathbf{A})$ one considers the following subgroup

$$G(\mathbf{A})^1 = \{x \in G(\mathbf{A}) : \| \det x \| = 1\}, \tag{4.1}$$

where $\| \det x\| = \prod_p |\det x|_p$. Then, $G(\mathbb{Q})$ is also embedded diagonally and discretely in $G(\mathbf{A})$. Moreover, it is a normal subgroup of $G(\mathbf{A})$ that can be also defined as follows.

Write $X(G)$ for the group of rational characters of G defined over $\mathbb{Q}$. Let

$$J = \mathrm{Hom}(X(G), \mathbb{R})$$
$$J^* = X(G) \otimes \mathbb{R}$$

Define a map

$$H_G : G(\mathbf{A}) \longrightarrow J$$

by

$$e^{\langle X, H_G(x)\rangle} = |X(x)|,$$

where $X \in X(G)$, $x \in G(\mathbf{A})$, and $\langle , \rangle$ is the pairing between J^* and J. Then,

$$G(\mathbf{A})^1 = \{x \in G(\mathbf{A}) : H_G(x) = 0\}. \tag{4.2}$$

The parabolic subgroups of $G(\mathbf{A})$, and $G(\mathbf{A})^1$ can be defined similarly as before, i.e., to the partition $(r_1, \ldots, r_\ell)$ of r we associate parabolic subgroups P, and we can write

$$P(\mathbf{A}) = M_P(\mathbf{A}) N_P(\mathbf{A}).$$

If $K_0 = \prod_{p<\infty} G(\mathbb{Z}_p)$, and $K = O(n, \mathbb{R}) \times K_0$, then

$$G(\mathbf{A}) = P(\mathbf{A})K.$$

Therefore an arbitrary element $x \in G(\mathbf{A})$ can be written as

$$x = mnk,$$

where $m \in M_P(\mathbf{A})$, $n \in N_P(\mathbf{A})$, $k \in K$.

Now, to x we associate the map H_P by:

$$H_P(x) = \left(\frac{1}{r_1} \log |\det m_1|, \ldots, \frac{1}{r_\ell} \log |\det m_\ell|\right) \in \mathfrak{a}_P \cong \mathbb{R}^\ell.$$

Note that there is a difference between this and our previous definition of $\mathfrak{a}_P$. In fact, let $P = G$, the difference is then transparent.

At this point our study can be pushed forward by relying on the theory of *Strong approximation.* According to this theory one knows that any element $x \in G(\mathbf{A})$ can be written as

$$x = \gamma x_\infty k_0,$$

where, $\gamma \in G(\mathbb{Q})$, $x_\infty \in G(\mathbb{R})$, and $k_0 \in K_0$. This means that the map

$$G(\mathbb{R}) \longrightarrow G(\mathbb{Q}) \backslash G(\mathbf{A}) / K_0$$

is surjective. From which one may extract that

$$G(\mathbb{Z}) \backslash G(\mathbb{R}) \cong G(\mathbb{Q}) \backslash G(\mathbf{A}) / K_0.$$

It is now clear that this isomorphism of topological spaces implies the isomorphism of the Hilbert spaces.

$$L^2(G(\mathbb{Z}) \backslash G(\mathbb{R})) \cong L^2(G(\mathbb{Q}) \backslash G(\mathbf{A}) / K_0), \tag{4.3}$$

and that $G(\mathbb{R})$ acts from the right on both sides. Observe that this latter isomorphism is a special case of a general one, and (4.3) is a consequence of the arithmetic of $\mathbb{Q}$, namely the fact that the class number of $\mathbb{Q}$ is one. To explain the general case, we appeal to a theorem of Borel ([26] . Sec. 5), (see also [28] . Sec. 4.7). To state this theorem recall that, if $\mathbf{A}_0$ is the ring of finite adèles, and K_0 an open compact subgroup of $G(\mathbf{A}_0)$, then Borel proves that:

Theorem (4.1) (A. Borel). $G(\mathbf{A})$ is the disjoint union of finitely many double cosets $G(\mathbb{Q})x_iG(\mathbb{R})K_0$, where $1 \leq i \leq c$, $x_i \in G(\mathbf{A}_0)$.

Put

$$\Gamma_i = G(\mathbb{Q}) \cap x_i G(\mathbb{R}) K_0 x_i^{-1} \quad i = 1, \ldots, c.$$

Then, Γ_i are arithmetic subgroups of $G(\mathbb{R})$, and

$$\coprod_{i=1} (\Gamma_i \backslash G(\mathbb{R})) \cong G(\mathbb{Q}) \backslash G(\mathbf{A}) / K_0.$$

The correspondence between functions on these spaces are described as follows. Let f be a function on $G(\mathbf{A})$ and x_i $(1 \leq i \leq c)$ one of the representatives in $G(\mathbf{A}_0)$. Then, define f_i on $G(\mathbb{R})$ by

$$f_i : x \longrightarrow f(x \cdot x_i).$$

Suppose that f is right-invariant under K_0. Then, f is left-invariant under $G(\mathbb{Q})$ if and only if f_i is left-invariant under Γ_i for every i $(1 \leq i \leq c)$.

Moreover, as a result of the above isomorphism we have

$$\bigoplus_{i=1}^{c} L_d^2(\Gamma_i \backslash G(\mathbb{R})) \cong L_d^2(G(\mathbb{Q}) \backslash G(\mathbf{A}) / K_0).$$

Here, observe that $G = GL(r, \mathbb{R})^1$, and thus the identity connected component of the center of $G(\mathbb{R})$ is reduced to the identity.

Now, we explain the choice of a "good" function on $G(\mathbf{A})$. The existence of this function was already pointed out in the previous chapters. The choice of these functions permit us to globalize our study. One choses f on $G(\mathbf{A})$ by taking it to be a finite sum of the functions $f_{\mathbb{R}} \otimes f_0$, where $f_{\mathbb{R}} \in C_c^\infty(G(\mathbb{R}))$, and f_0 is a function on $G(\mathbf{A}_0)$ such that, (i) $f_{\mathbb{R}}$ is compactly supported, (ii) f_0 is locally constant. The space of the functions f is denoted by $C_c^\infty(G(\mathbf{A}))$. For example, if $f = f_{\mathbb{R}} \otimes f_0$, and $x \in G(\mathbf{A})$, then

$$f(x) = f_{\mathbb{R}}(x_{\mathbb{R}}) f_0(x_0),$$

where, $x = x_{\mathbb{R}} x_0$ is the product of the Archimedean component $x_{\mathbb{R}}$, and the non-Archimedean component x_0.

We end this section by recalling that Reduction theory can be put in the frame of adèlic language. This was already done by Godement [46], (see also [82]). For example, the analogue of the Siegel domain can similarly be defined in an adèlic setting. To do this, let $P_0(\mathbf{A})$ be the minimal parabolic subgroup corresponding to the partition $(1, \ldots, 1)$. Then

$$M_{P_0}(\mathbf{A}) \cong M_{P_0}(\mathbb{Q}) \cdot A_{P_0}(\mathbb{R})^0 \cdot (M_{P_0}(\mathbf{A}) \cap K),$$

where $A_{P_0}(\mathbb{R})^0$ is the identity connected component of $A_{P_0}(\mathbb{R})$. However, we have in general

$$M_P(\mathbf{A}) = M_P(\mathbf{A})^1 A_P(\mathbb{R})^0,$$

where $M_P(\mathbf{A})^1$ can be defined similarly as in the real case by replacing the absolute value by the idele norm, or equivalently noticing that $M(\mathbf{A})^1$ is the kernel of the homomorphism H_M to the additive group $\mathbf{a}_P$. Thus

$$G(\mathbf{A}) = N(\mathbf{A})M(\mathbf{A})^1 A_P(\mathbb{R})^0 K.$$

Exercise (4.1). Prove that

$$M_{P_0}(\mathbf{A}) \cong M_{P_0}(\mathbb{Q}) \cdot A_{P_0}(\mathbb{R})^0 \cdot (M_{P_0}(\mathbf{A}) \cap K).$$

Hint: observe that the group of ideles $\mathbf{A}^*$ can be expressed as

$$\mathbf{A}^* = \mathbb{Q}^* \cdot (\mathbb{R}_+^* \times \prod_{P<\infty} \mathbb{Z}_P^*).$$

For a proof of this see ([115], ch. IV. Th. 7, and its corollary).

Let us now define the adèlic Siegel domain.

Let $T_0 \in \mathbf{a}_{P_0}$, and w be a compact subset of $N_{P_0}(\mathbf{A})$. Then, the *adèlic Siegel domain* is denoted by $S(T_0, w)$ and it is defined by:

$$\{nak : n \in w,\ k \in K,\ a \in A_P(\mathbb{R})^0,\ \alpha(H_{P_0}(a) - T_0) > 0 \quad \forall \alpha \in \Delta_{P_0}\}.$$

In particular one can show that

$$G(\mathbf{A}) = G(\mathbb{Q}) S(T_0, w).$$

Chapter V

The Geometric Theory

In this chapter we study three basic notions in order to understand the geometric expansion of Arthur's trace formula. They are discussed in three sections, their definitions are the out come of the attempts to make the trace formula invariant, and as we shall see, they are related to the geometry of certain convex polytopes.

5.1. The $J_o^T(f)$ and $J^T(f)$ distributions

The purpose of this section is to introduce the two fundamental distributions, $J_o^T(f)$, and $J^T(f)$. These are polynomials in T, and they are used in the construction of the geometric expansion of the trace formula. To define these distributions we need first to introduce certain equivalence relation and conjugacy classes in $G(\mathbb{Q}) = GL(r, \mathbb{Q})$, and then, to define certain kernel functions parametrized by these classes.

Let $\gamma \in G(\mathbb{Q}) = GL(r, \mathbb{Q})$; define the *Jordan decomposition* of γ by

$$\gamma = \gamma_s \gamma_u,$$

where, $\gamma_u \in G(\mathbb{Q})$ is semi-simple, and $\gamma_s \in G(\mathbb{Q})$ is unipotent. One knows that these terms are unique and commute with each other. Now, define a relation between two elements $\gamma, \gamma' \in G(\mathbb{Q})$, by demanding that the semi-simple parts of γ and γ' be conjugate in $G(\mathbb{Q})$. Thus,

$$\gamma \sim \gamma' \Leftrightarrow \gamma_s \text{ is conjugate to } \gamma'_s.$$

Let $\mathcal{O}$ be the set of all equivalence classes in $G(\mathbb{Q})$.

Example (5.1). (i) In $G_2 = GL(2)$, the element $\begin{pmatrix} 0 & 1 \\ 1 & a \end{pmatrix}$, can be written as the product

$$\begin{pmatrix} 0 & 1 \\ 1 & a \end{pmatrix} = \begin{pmatrix} 0 & 1 \\ 1 & 0 \end{pmatrix} \begin{pmatrix} 1 & a \\ 0 & 1 \end{pmatrix},$$

in which, $\begin{pmatrix} 0 & 1 \\ 1 & 0 \end{pmatrix}$ is semi-simple (its characteristic polynomial is the product of linear factors), and $\begin{pmatrix} 1 & a \\ 0 & 1 \end{pmatrix}$ is unipotent.

(ii) Suppose that $o \in \mathcal{O}$. Then the characteristic polynomial of o is the product of linear factors (i.e., all the eigenvalues are different) if and only if $o \cap P = \emptyset$ for every parabolic subgroup $P \neq G$.

(iii) For any parabolic subgroup P, and any $o \in \mathcal{O}$

$$o \cap P(\mathbb{Q}) = (o \cap M_P(\mathbb{Q}))N_P(\mathbb{Q}).$$

Now we have enough information to define several kernel functions and the $J^T(f)$ and $J_0^T(f)$ distributions.

Recall that, since we are adopting the adèlic language, the analogue of $\Gamma\backslash G$ will be $G(\mathbf{Q})\backslash G(\mathbf{A})^1$. The first kernel function is the one of Selberg, where Γ is replaced by $G(\mathbf{Q})$, i.e.,

$$\begin{aligned}\text{(i)}\quad K(x,x) &= \sum_{\gamma\in G(\mathbf{Q})} f(x^{-1}\gamma x) \quad (f\in C_c^\infty(G(\mathbf{A}))\\ &= \sum_{o\in\mathcal{O}}\sum_{\gamma\in o} f(x^{-1}\gamma x),\end{aligned}$$

so that we write the *partial kernel function* as

$$\text{(ii)}\quad K_o(x,x) = \sum_{\gamma\in o} f(x^{-1}\gamma x).$$

$$\begin{aligned}\text{(iii)}\quad K_P(x,x) &= \sum_{\gamma\in M_P(\mathbf{Q})}\int_{N_P(\mathbf{A})} f(x^{-1}\gamma n x)dn\\ &= \sum_{o\in\mathcal{O}} K_{P,o}(x,x),\end{aligned}$$

where the *partial kernel function* in this case is given by

$$\text{(iv)}\quad K_{P,o}(x,x) = \sum_{\gamma\in M_P(\mathbf{Q})\cap o}\int_{N_P(\mathbf{A})} f(x^{-1}\gamma n x)dn,$$

when $M_P(\mathbf{Q})\cap o = \phi$, otherwise, we set $K_{P,o}(x,x)=0$.

(v) Define the *partial Arthur's kernel function* by

$$k_o^T(x,f) = \sum_P (-1)^{\dim(A_P/A_G)} \sum_{\delta\in P(\mathbf{Q})\backslash G(\mathbf{Q})} K_{P,o}(\delta x,\delta x)\hat{\tau}_P(H_P(\delta x)-T_P),$$

where, as before $T_P=(t_1,\ldots,t_r)$ is such that $t_i \gg t_{i+1}$. Then Arthur's kernel function is the summation of the partial kernel functions given below.

$$\text{(vi)}\quad k^T(x,f) = \sum_{o\in\mathcal{O}} k_o^T(x,f).$$

Now we define the distributions

$$\begin{aligned} J^T(f) &= \int_{G(\mathbf{Q})\backslash G(\mathbf{A})^1} k^T(x,f)dx\\ &= \int_{G(\mathbf{Q})\backslash G(\mathbf{A})^1} \sum_o k_o^T(x,f)dx\\ &= \sum_o \int_{G(\mathbf{Q})\backslash G(\mathbf{A})^1} k_o^T(x,f)dx,\end{aligned}$$

and

$$J_o^T(f) = \int_{G(\mathbf{Q})\backslash G(\mathbf{A})^1} k_o^T(x,f)dx.$$

Hence

$$J^T(f) = \sum_{o\in\mathcal{O}} J_o^T(f).$$

To clarify these notations, let us consider a special case.

Recall that a *distribution* on $C_c^\infty(G(\mathbf{A}))$ is a linear function on this space. Thus, $J_{\mathfrak{o}}^T(f)$, and $J^T(f)$ are both distributions.

Example (5.2). Suppose $\mathfrak{o} \cap P = \emptyset \quad \forall \ P \neq G$, then $\mathfrak{o} = \{\gamma_1\}$ is a semisimple class in $G(\mathbf{Q})$. We calculate $J_{\mathfrak{o}}^T(f)$. This amounts to calculating the kernel $k_{\mathfrak{o}}^T(x, f)$ first. Since our only parabolic subgroup which intersects $\mathfrak{o}$ is G, and since by definition $\hat{\tau}_G \equiv 1$, we get

$$\begin{aligned} k_{\mathfrak{o}}^T(x,f) &= K_{P,\mathfrak{o}}(x,x) \\ &= \sum_{\gamma\in\mathfrak{o}} f(x^{-1}\gamma x) \\ &= K_{\mathfrak{o}}(x,x). \end{aligned}$$

Hence

$$\begin{aligned} J_{\mathfrak{o}}^T(f) &= \int_{G(\mathbf{Q})\backslash G(\mathbf{A})^1} \sum_{\gamma\in\{\gamma_1\}} f(x^{-1}\gamma x)dx \\ &= \int_{G(\mathbf{Q})\backslash G(\mathbf{A})^1} \sum_{\delta\in G_{\gamma_1}(\mathbf{Q})\backslash G(\mathbf{Q})} f(x^{-1}\delta^{-1}\gamma_1\delta x)dx \\ &= \mathrm{vol}(G_{\gamma_1}(\mathbf{Q})\backslash G_{\gamma_1}(\mathbf{A})^1)\int_{G_{\gamma_1}(\mathbf{A})^1\backslash G(\mathbf{A})^1} f(x^{-1}\gamma_1 x)dx. \end{aligned}$$

Observe that the last expression for $J_{\mathfrak{o}}^T(f)$ is the same as the geometric expansion of Selberg's trace formula. This is of course due to our condition on $\mathfrak{o}$. As a result, we have also proved the following.

Theorem (5.3). If H is a reductive group such that the quotient space $H(\mathbf{Q})\backslash H(\mathbf{A})^1$ is compact, then $J_{\mathfrak{o}}^T(f)$ is the geometric expansion of Selberg's trace formula. In another words, $J_{\mathfrak{o}}^T(f)$ generalizes the geometric expansion of Selberg's trace formula.

In fact, when the quotient space $H(\mathbf{Q})\backslash H(\mathbf{A})^1$ is non-compact, $H(\mathbf{Q})$ contains parabolic subgroups outside of its radical [82], and the more parabolic subgroups intersect $\mathfrak{o}$, the more complicated $J_{\mathfrak{o}}^T(f)$ will be.

In the following section by using the geometric Γ−function of Arthur we show that $J_{\mathfrak{o}}^T(f)$ is a polynomial in T.

5.2. A geometric Γ−function

The purpose of this section is two-fold. First, we define the geometric Γ−function of Arthur, which is the key to the understanding of the geometric expansion of the trace formula. Secondly, we use this function to prove that the distributions $J_{\mathfrak{o}}^T(f)$ and $J^T(f)$ are polynomials in T.

Fix T_1 and consider the difference

$$\hat{\tau}_P(H_P(\delta x) - T) - \hat{\tau}_P(H_P(\delta x) - T_1).$$

Set

$$H_P(\delta x) - T = H - X, \quad H_P(\delta x) - T_1 = H,$$

so

$$X = T - T_1.$$

Suppose that H and X are two points in $\mathfrak{a}_{P_0}$. Then, for every P, define the Γ*-function* $\Gamma_P(H,X)$ inductively, by first supposing that it is defined for any parabolic subgroup $P \subseteq Q \subsetneq G$, then define it for Q by:

$$\hat{\tau}_P(H-X) = \sum_{\{Q:Q\supset P\}} (-1)^{\dim(A_Q/A_G)} \hat{\tau}_P^Q(H)\Gamma_Q(H,X).$$

Whenever $Q = G$, define

$$\Gamma_G(H,X) = 1.$$

The following lemma is fundamental for the study of $J_{\mathfrak{O}}^T$.

Lemma (5.4). If H is restricted to $\mathfrak{a}_P^G$, then $\Gamma_P(H,X)$ is compactly supported on H, and $\int_{\mathfrak{a}_P^G} \Gamma_P(H,X)dH$ is a polynomial in X.

Proof. [4] Lemmas 2.1 and 2.2. ∎

Example (5.5). Let $r = 3$. We now have two cases.

(a) $P_1 = P$, so $\mathfrak{a}_P/\mathfrak{a}_G$ is one-dimensional, therefore $\Gamma_P(H,X)$, $H \in \mathfrak{a}_P^G$ is the characteristic function of the bounded shaded interval

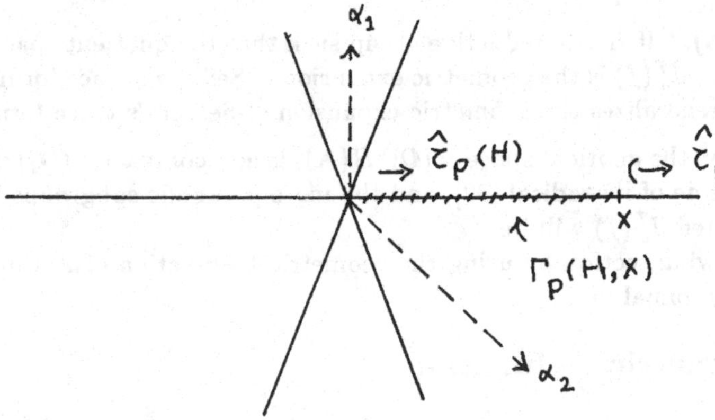

(b) $P = P_0$, so $\mathfrak{a}_{P_0}/\mathfrak{a}_G$ is two-dimensional. In this case, $\Gamma_{P_0}(H,X)$, $H \in \mathfrak{a}_{P_0}^G$ is the characteristic function of the bounded shaded region.

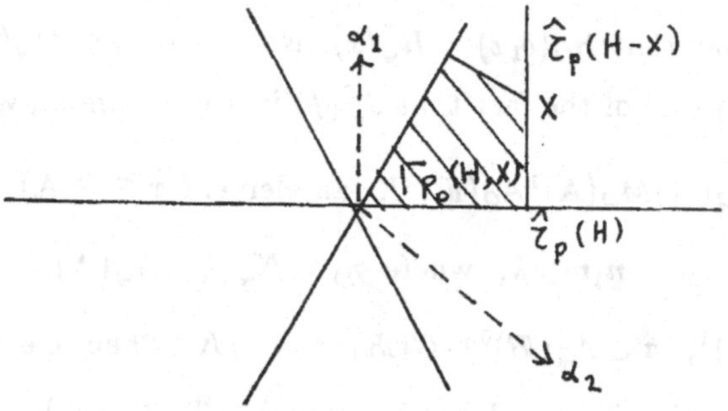

Let us return to our discussion of $J_o^T(f)$. In the expression for $J_o^T(f)$, make the substitution

$$\sum_{\{Q:Q\supset P\}} (-1)^{\dim(A_Q/A_G)}\hat{\tau}_P^Q(H_P(\delta x)-T_1)\Gamma_Q(H_Q(\delta x)-T_1,T-T_1),$$

for $\hat{\tau}_P(H_P(\delta x)-T)$.

Hence we get

$$J_o^T(f)=\int_{G(\mathbf{Q})\backslash G(\mathbf{A})^1}\sum_P(-1)^{\dim(A_P/A_G)}\sum_{\delta\in P(\mathbf{Q})\backslash G(\mathbf{Q})}\sum_{Q\supset P}(-1)^{\dim(A_Q/A_G)}\times$$
$$K_{P,o}(\delta x,\delta x)\hat{\tau}_P^Q(H_P(\delta x)-T_1)\Gamma_Q(H_Q(\delta x)-T_1,T-T_1).$$

Take the sum over Q outside the sum over P, then write the sum over $\delta\in P(\mathbf{Q})\backslash G(\mathbf{Q})$ as an integral over $P(\mathbf{Q})\backslash G(\mathbf{Q})$. Now write the integral over $(G(\mathbf{Q})\backslash G(\mathbf{A})^1)\times(P(\mathbf{Q})\backslash G(\mathbf{Q}))$ as an integral over $(Q(\mathbf{Q})\backslash G(\mathbf{A})^1)\times(P(\mathbf{Q})\backslash Q(\mathbf{Q}))$, and write the integral over $P(\mathbf{Q})\backslash Q(\mathbf{Q})$ as a sum over $\delta\in P(\mathbf{Q})\backslash Q(\mathbf{Q})$, to get

$$J_o^T(f) = \sum_Q\int_{Q(\mathbf{Q})\backslash G(\mathbf{A})^1}\sum_{P\subset Q}(-1)^{\dim(A_P/A_G)}\sum_{\delta_1\in P(\mathbf{Q})\backslash Q(\mathbf{Q})}$$
$$K_{P,o}(\delta_1 x,\delta_1 x)\hat{\tau}_P^Q(H_P(\delta_1 x)-T_1)\Gamma_Q(H_Q(\delta_1 x)-T_1,T-T_1).$$

Note that by the isomorphism theorem, we have

$$P(\mathbf{Q})\backslash Q(\mathbf{Q})\cong P(\mathbf{Q})\cap M_Q(\mathbf{Q})\backslash M_Q(\mathbf{Q}),$$

then apply this to the summation over δ_1 above to obtain:

$$J_o^T(f) = \sum_Q\int_{Q(\mathbf{Q})\backslash G(\mathbf{A})^1}\sum_{P\subset Q}(-1)^{\dim(A_P/A_G)}\sum_{\delta_1\in P(\mathbf{Q})\cap M_Q(\mathbf{Q})\backslash M_Q(\mathbf{Q})}$$
$$K_{P,o}(\delta_1 x,\delta_1 x)\hat{\tau}_P^Q(H_P(\delta_1 x)-T_1)\Gamma_Q(H_Q(x)-T_1,T-T_1)dx.$$

The changes of summations and integrals in the above computations are justified by the convergence of integrals and the finiteness of the summations. Moreover, in our last computation for the final form of $J_o^T(f)$ we have applied the following exercise.

Exercise (5.6). Show that $H_Q(\delta_1 x) = H_Q(x)$. Note that $\delta_1 \in M_Q(\mathbf{Q})$.

To complete the proof of the fact that $J_o^T(f)$ is a polynomial, we need to carry out some calculations.

Since $G(\mathbf{A}) = N_Q(\mathbf{A})M_Q(\mathbf{A})^1 A_Q(I\!R)^0 K$, an element $x \in G(\mathbf{A})$ can be written as

$$x = n_1 m_1 a k, \quad \text{where } n_1 \in N_Q(\mathbf{Q})\backslash N_Q(\mathbf{A}),$$

$m_1 \in M_Q(\mathbf{Q})\backslash M_Q(\mathbf{A})^1$, $a \in A_Q(I\!R)^0 \cap G(I\!R)^1$, $k \in K$. Then one can verify that

(i) $\Gamma_Q(H_Q(x) - T_1, T - T_1) = \Gamma_Q(H_Q(a) - T_1, T - T_1)$,

(ii) $\hat{\tau}_P^Q(H_P(\delta_1 x) - T_1) = \hat{\tau}_P^Q(H_P(\delta_1 m_1) - T_1)$.

Therefore

$$\delta_Q(a)K_{P,o}(\delta_1 x, \delta_1 x) = \delta_Q(a)\int_{N_P(\mathbf{A})} \sum_{\gamma \in o \cap M_P(\mathbf{Q})} \times$$

$$f(k^{-1}m_1^{-1}a^{-1}m_1^{-1}\delta_1^{-1}\gamma n \delta_1 a m_1 k)dn$$

$$= \int_{N_P(\mathbf{A})} \sum_{\gamma \in o \cap M_P(\mathbf{Q})} f(k^{-1}m_1^{-1}\delta_1^{-1}\gamma\delta_1 m_1 n k)dn.$$

But, $N_P(\mathbf{A}) = N_Q(\mathbf{A})N_P^Q(\mathbf{A})$, where $N_P^Q = N_P \cap M_Q$ if $Q \supset P$, (as is the case here). Hence the integral above can be written as

$$\int_{N_Q(\mathbf{A})N_P^Q(\mathbf{A})} \sum_{\gamma \in o \cap M_P(\mathbf{Q})} f(k^{-1}m_1^{-1}\delta_1^{-1}\gamma\delta_1 m_1 n k)dn.$$

This is the only term which depends on k, and the integral over k equals

$$\int_{N_P^Q(\mathbf{A})} \sum_{\gamma \in o \cap M_P(\mathbf{Q})} \int_K \int_{N_Q(\mathbf{A})} f(k^{-1}m_1^{-1}\delta_1^{-1}\gamma\delta_1 m_1 n k)dn$$

$$= \int_{N_P^Q(\mathbf{A})} \sum_{\gamma \in o \cap M_P(\mathbf{Q})} f_Q(m_1^{-1}\delta_1^{-1}\gamma\delta_1 m_1)dn,$$

where $f_Q(\mu) = \delta_Q(\mu)^{\frac{1}{2}} \int_K \int_{N_Q(\mathbf{A})} f(k^{-1}\mu n k)dndk$, $\mu \in M_Q(\mathbf{A})$. We note that the kernel of δ_Q contains $M_Q(\mathbf{A})^1$, from which it turns out that $\delta_Q(\mu) = e^{2\rho(H_Q(\mu))}$, where $2\rho =$ (sum of positive roots of (Q, A_Q)). See also [3], page 918.

Observe that the definition of f_Q implies that $f_Q \in C_c^\infty(M_Q(\mathbf{A}))$. Therefore the integral above over $N_P^Q(\mathbf{A})$ equals $K^{M_Q}_{P\cap M_Q, o\cap M_Q}(\delta_1 m_1, \delta_1 m_1)$, where $K^{M_Q}_{P\cap M_Q, o\cap M_Q}$ denotes the kernel $K_{P,Q}$, but with (G, P, f, o) replaced by $(M_Q, P \cap M_Q, f_Q, o \cap M_Q(\mathbf{Q}))$. Hence

$$J_o^T(f) = \sum_{\{Q: Q \supset P\}} \int_{M_Q(\mathbf{Q})\backslash M_Q(\mathbf{A})^1} \left(\sum_{\{P: P \subset Q\}} (-1)^{\dim(A_{P\cap M_Q}/A_{M_Q})} \right.$$

$$\sum_{\delta_1\in P(Q)\cap M_Q(Q)\backslash M_Q(Q)} K^{M_Q}_{P\cap M_Q,o}(\delta_1 m_1,\delta_1 m_1)\hat{\tau}^Q_P(H_P(\delta_1 m_1)-T_1)dm_1\times$$

$$\left.\int_{A_Q(\mathbb{R})^0\cap G(\mathbb{R})^1}\Gamma_Q(H_Q(a)-T_1,T-T_1)da\right).$$

Finally, write $H_Q(a)=H$, and note that $A_Q(\mathbb{R})^0\cap G(\mathbb{R})^1\cong \mathbf{a}^G_Q$, then

$$J^T_o(f)=\sum_{Q\supset P} J^{M_Q,T}_{o\cap M_Q}(f_Q)\int_{\mathbf{a}^G_Q}\Gamma_Q(H,T-T_1)dH.$$

Hence we have proved that $J^T_o(f)$ is a polynomial in T, since the right-hand side of the above expression is a polynomial in $T-T_1$.

We have thus proved the following theorem.

Theorem (5.7). The distributions $J^T_o(f)$ and $J^T(f)$ are polynomials in T.

Remark (5.8). (a) The degree of the polynomial of $J^T_o(f)$ is equal to $\dim(A_Q/A_G)$, where Q is the parabolic subgroup which is the minimal group among those that satisfy

$$Q\cap o\neq\emptyset\Leftrightarrow M_Q\cap o\neq\emptyset.$$

(b) $J^T_o(f)$ was originally defined only for T which was sufficiently regular in open sets in $\mathbf{a}_{P_0}$. In fact $J^T_o(f)$ can be defined for all $T\in\mathbf{a}_{P_0}$.

5.3. The Weight functions

We have already defined the notion of invariant distributions, and as an example, we saw that orbital integrals

$$f_{G(\mathbf{A})}(\gamma)=\int_{G_\gamma(\mathbf{A})\backslash G(\mathbf{A})} f(x^{-1}\gamma x)dx$$

are invariant distributions. We have also seen in Example (5.2), that under certain restriction, the distribution $J^T_o(f)$ is invariant, being a multiple of an orbital integral. We now consider the question, of whether in general $J^T_o(f)$ or $J^T(f)$ is an invariant distribution. More especifically, let us set $T=0$ in the polynomial $J^T_o(f)$ and define

$$J_o(f)=J^T_o(f)|_{T=0}.$$

Likewise we set

$$J(f)=J^T(f)|_{T=0}.$$

We then ask:

Question. For an arbitrary $o\in\mathcal{O}$, is J_o an invariant distribution?

To answer such a question, one has to look at the difference

$$J^T_o(f^y)-J^T_o(f),$$

where $f^y(x)=f(yxy^{-1})$. If the difference is zero then the distribution is invariant.

The answer to the above question is generally no, as we shall see below.

$J_o^T(f^y)$ has the same formula as $J_o^T(f)$, except that $K_{P,o}(\delta x,\delta x)$ is replaced by

$$\sum_{\gamma\in M_P(Q)\cap o}\int_{N_P(\mathbf{A})} f(yx^{-1}\delta^{-1}\gamma n\delta xy^{-1})dn = K_{P,o}(\delta xy^{-1},\delta xy^{-1}),$$

i.e.,

$$J_o^T(f^y) = \int_{G(Q)\backslash G(\mathbf{A})^1}\sum_P(-1)^{\dim(A_P/A_G)}\sum_{\delta\in P(Q)\backslash G(Q)} K_{P,o}(\delta xy^{-1},\delta xy^{-1})\hat{\tau}_P(H_P(\delta x)-T)dx.$$

Changing the variable x to xy one gets:

$$J_o^T(f^y) = \int_{G(Q)\backslash G(\mathbf{A})^1} dx\left(\sum_P(-1)^{\dim(A_P/A_G)}\times \sum_{\delta\in P(Q)\backslash G(Q)} K_P(\delta x,\delta x)\hat{\tau}_P(H_P((\delta xy)-T)\right).$$

If $\delta x = nmak$, where $n\in N_P(\mathbf{A})$, $m\in M_P(\mathbf{A})^1$, $a\in A_P(\mathbb{R})^0$, $k\in K$, let $K_P(\delta x)$ be any element in K such that $(\delta x\cdot K_P(\delta x)^{-1})\in P(\mathbf{A})$. Then write $k'=K_P(\delta x)$, from which we get:

$$\begin{aligned}\hat{\tau}_P(H_P(\delta xy)-T) &= \hat{\tau}_P(H_P(a)+H_P(k'y)-T))\\ &= \hat{\tau}_P(H_P(\delta x)-T+H_P(K_P(\delta x)y)\\ &= \sum_{\{Q:Q\supset P\}}(-1)^{\dim(A_Q/A_G)}\times\\ &\qquad \hat{\tau}_P^Q(H_P(\delta x)-T)\Gamma_Q(H_P(\delta x)-T,-H_P(K_P(\delta x)y)).\end{aligned}$$

Observe that

$$\Gamma_Q(H_P(\delta x)-T,\ -H_P(K_P(\delta x)y)) = \Gamma_Q(H_Q(\delta x)-T,\ H_Q(K_Q(\delta x)y)).$$

Now define,

$$u_Q(k,y) = \int_{\mathfrak{a}_Q^G}\Gamma_Q(H,-H_Q(ky))dH,\qquad k\in K,$$

and set $H=H_Q(a)$ as before. We get:

$$J_o^T(f^y) = \sum_Q J_{o\cap M_Q}^{M_Q,T}(f_{Q,y}),$$

where

$$f_{Q,y}(m) = \delta_Q(m)^{\frac{1}{2}}\int_K\int_{N_Q(\mathbf{A})} f(k^{-1}mnk)u_Q(k,y)dndk,$$

$m\in M_Q(\mathbf{A})$, $f_{Q,y}\in C_c^\infty(M_Q(\mathbf{A}))$. Set $T=0$, and obtain:

$$J_o(f^y) = \sum_o J_{o\cap M_Q}^{M_Q}(f_{Q,y}). \tag{5.1}$$

Note that, by definition, $f_{G,y} = f$. From this it follows a very important result which shows how much $J_o(f)$ is non-invariant. This degree of non-invariance is

$$J_o(f^y) - J_o(f) = \sum_{Q \neq G} J^{M_Q}_{o \cap M_Q}(f_{Q,y}).$$

It is easy to see that the right-hand side of the equality above is a distribution, and our aim is to write it in terms of orbital integrals.

Remark. One expects that when D is an invariant distribution on $G(\mathbb{R})$, then $D(f)$ depends only on the function

$$g(\gamma) = \Delta(\gamma) \int_{G_\gamma(\mathbb{R}) \backslash G(\mathbb{R})} f(x^{-1}\gamma x) dx,$$

where $\gamma \in G\text{reg}(\mathbb{R})$, and Δ is a certain function of γ. Or alternatively D depends only on the functioin $h(\pi) = \text{tr}\ \pi(f)$ with $\pi \in G(\hat{\mathbb{R}})$.

We now suggest a problem.

Problem (5.9). Find a more explicit formula for $J_o(f)$.

From this point on, we shall devote a large part of these notes to obtain a satisfactory solution to the problem above, so that our explicit formula for $J_o(f)$ will be given as a weighted orbital integral, which in fact reduces to an ordinary orbital integral when there is no unipotent conjugacy classes. Hence our answer (again) generalizes Selberg's trace formula, as a natural consequence.

Lemma (5.10). Let $\gamma \in M_P(\mathbb{Q})$, $\phi \in C_c^\infty(N_P(\mathbf{A}))$, $\gamma = \gamma_s \gamma_u$ be the Jordan decomposition, and let N_{P,γ_s}, be the centralizer of γ_s in N_P, then

$$\sum_{\gamma \in N_P(\mathbb{Q})} \phi(\gamma) = \sum_{\eta \in N_{P,\gamma_s}(\mathbb{Q}) \backslash N_P(\mathbb{Q})} \sum_{\tau \in N_{P,\gamma_s}(\mathbb{Q})} \phi(\gamma^{-1}\eta^{-1}\gamma\tau\eta),$$

and, moreover,

$$\int_{N_P(\mathbf{A})} \phi(n) dn = \int_{N_{P,\gamma_s}(\mathbf{A}) \backslash N_P(\mathbf{A})} \int_{N_{P,\gamma_s}(\mathbf{A})} \phi(\gamma^{-1}u^{-1}\gamma n_1 u) dn_1 du.$$

Proof. See [3], Lemma 2.1. ■

We now define,

$$\tilde{K}_{P,o}(x,y) = \sum_{\gamma \in M_P(\mathbb{Q}) \cap o} \sum_{\eta \in N_{P,\gamma_s}(\mathbb{Q}) \backslash N_P(\mathbb{Q})} \sum_{N_{P,\gamma_s}(\mathbf{A})} f(x^{-1}\eta^{-1}\gamma n \eta y) dn.$$

Let us also define,

$$\tilde{K}^T_o(x,f) = \sum_P (-1)^{dim(A_P/A_G)} \sum_{\delta \in P(\mathbb{Q}) \backslash G(\mathbb{Q})} \tilde{K}_{P,o}(\delta x, \delta x) \hat{\tau}_P(H_P(\delta x) - T).$$

Exercise (5.11). (a) Show that when $\gamma_s = \{1\}$ i.e. o consists only of the unipotent classes, then

$$\sum_{\eta \in N_{P,\gamma_s}(\mathbb{Q}) \backslash N_P(\mathbb{Q})} \int_{N_{P,\gamma_s}(\mathbf{A})} f(x^{-1}\eta^{-1}\gamma n \eta y) dn = 0.$$

Hence $\tilde{k}_o^T(x,f) = k_o^T(x,f)$.

(b) If o consists entirely of semi-simple elements, then

$$\tilde{K}_P(x,y) = \sum_{\gamma\in M_P(\mathbf{Q})\cap o}\ \sum_{\eta\in N_P(\mathbf{Q})} f(x^{-1}\eta^{-1}\gamma\eta y).$$

Theorem (5.12). Here we have a variation of theorem (3.17) for $\tilde{k}_o^T(x,f)$ i.e.,

$$\int_{G(\mathbf{Q})\backslash G(\mathbf{A})^1} |\tilde{k}_o^T(x,f)|dx < \infty.$$

The proof of this theorem is the same as for Theorem (3.17). However, here we prove that

$$\int_{G(\mathbf{Q})\backslash G(\mathbf{A})^1} \tilde{k}_o^T(x,f)dx = \int_{G(\mathbf{Q})\backslash G(\mathbf{A})^1} k_o^T(x,f)dx \tag{5.2}$$

when T is suitably regular. Recall that our convergence theorem in the adèlic case gives

$$\int_{G(\mathbf{Q})\backslash G(\mathbf{A})^1} k_o^T(x,f)dx = \sum_{P_1\subset P_2} \int_{P_1(\mathbf{Q})\backslash G(\mathbf{A})^1} A(x)B(x)dx,$$

Where $A(x) = F^{P_1}(x,T)\sigma_{P_1}^{P_2}(H_{P_1}(x) - T)$,

$$B(x) = \sum_{\{P:P_1\subset P\subset P_2\}} (-1)^{\dim(A_P/A_G)} \sum_{\gamma\in M_P(\mathbf{Q})\cap o} \int_{N_P(\mathbf{A})} f(x^{-1}\gamma n x)dn.$$

Similarly

$$\int_{G(\mathbf{Q})\backslash G(\mathbf{A})^1} \tilde{k}_o^T(x,f)dx = \sum_{P_1\subset P_2} \int_{P(\mathbf{Q})\backslash G(\mathbf{A})^1} A(x)\tilde{B}(x)dx,$$

where

$$\tilde{B}(x) = \sum_{\{P:P_1\subset P\subset P_2\}} (-1)^{\dim(A_P/A_G)} \sum_{\gamma\in M_P(\mathbf{Q})\cap o}\ \sum_{\eta\in N_{P,\gamma_s}(\mathbf{Q})\backslash N_P(\mathbf{Q})}$$

$$\int_{N_{P,\gamma_s}(\mathbf{A})} f(x^{-1}\eta^{-1}\gamma n\eta x)dn.$$

Then

$$\int_{P_1(\mathbf{Q})\backslash G(\mathbf{A})^1} A(x)\tilde{B}(x)dx =$$

$$\int_{M_{P_1}(\mathbf{Q})N_{P_1}(\mathbf{A})\backslash G(\mathbf{A})^1} \int_{N_{P_1}(\mathbf{Q})\backslash N_{P_1}(\mathbf{A})} A(n_1x)\tilde{B}(n_1x)dn_1dx,$$

in fact $A(n_1x) = A(x)$ since $A(x)$ is invariant from the left under $N_{P_1}(\mathbf{A})$.

To calculate

$$\int_{N_{P_1}(\mathbf{Q})\backslash N_{P_1}(\mathbf{A})} \tilde{B}(n_1x)dn_1,$$

take

$$\int_{N_{P_1}(\mathbf{Q})\backslash N_{P_1}(\mathbf{A})} dn_1 \text{ inside } \sum_P, \sum_\gamma \text{ i.e.,}$$

$$\int_{N_{P_1}(\mathbb{Q})\backslash N_{P_1}(\mathbb{A})} \tilde{B}(n_1x)dn_1 = \sum_{\{P:P_1\subset P\subset P_2\}} \sum_{\gamma\in M_P(\mathbb{Q})\cap o} (-1)^{\dim(A_P/A_G)} \times$$

$$\int_{N_{P_1}(\mathbb{Q})\backslash N_{P_1}(\mathbb{A})} dn_1 \sum_{\eta\in N_{P_1,\gamma_s}(\mathbb{Q})\backslash N_{P_1}(\mathbb{Q})} \int_{N_{P_1,\gamma_s}(\mathbb{A})} f(x^{-1}n^{-1}\eta^{-1}\gamma n\eta n_1 x)dn.$$

which is equal to

$$\sum_{\{P:P_1\subset P\subset P_2\}} \sum_{\gamma\in M_P(\mathbb{Q})\cap o} (-1)^{\dim(A_P/A_G)} \times$$

$$\int_{N_{P_1}(\mathbb{Q})\backslash N_{P_1}(\mathbb{A})} dn_1 \int_{N_{P,\gamma_s}(\mathbb{A})\backslash N_P(\mathbb{A})} d\nu \times$$

$$\int_{N_{P,\gamma_s}(\mathbb{A})} f(x^{-1}n^{-1}\nu^{-1}\gamma n\nu n_1 x)dn.$$

And this is equal to

$$\sum_{\{P:P_1\subset P\subset P_2\}} \sum_{\gamma\in M_P(\mathbb{Q})\cap o} (-1)^{\dim(A_P/A_G)} \times$$

$$\int_{N_{P_1}(\mathbb{Q})\backslash N_{P_1}(\mathbb{A})} dn_1 \int_{N_P(\mathbb{A})} f(x^{-1}n_1^{-1}\gamma_1 x)dn.$$

Taking $\int_{N_{P_1}(\mathbb{Q})\backslash N_{P_1}(\mathbb{A})} dn_1$ back outside the sums over γ and P, one gets:

$$\int_{N_{P_1}(\mathbb{Q})\backslash N_{P_1}(\mathbb{A})} \tilde{B}(n_1x)dn_1.$$

Therefore, we have

$$\int_{N_{P_1}(\mathbb{Q})N_{P_1}(\mathbb{A})\backslash G(\mathbb{A})^1} \int_{N_{P_1}(\mathbb{Q})\backslash N_{P_1}(\mathbb{A})} A(n_1x)\tilde{B}(n_1x)dn_1dx$$

$$= \int_{N_{P_1}(\mathbb{Q})N_{P_1}(\mathbb{A})\backslash G(\mathbb{A})^1} dx \int_{N_{P_1}(\mathbb{Q})\backslash N_{P_1}(\mathbb{A})} A(n_1x)B(n_1x)dn_1$$

$$= \int_{P_1(\mathbb{Q})\backslash G(\mathbb{A})^1} A(x)B(x)dx.$$

This implies,

$$\int_{G(\mathbb{Q})\backslash G(\mathbb{A})^1} \tilde{k}_o^T(x,f)dx = \int_{G(\mathbb{Q})\backslash G(\mathbb{A})^1} k_o^T(x,f)dx.$$

Hence we have proved (5.2).

Clearly the kernel

$$\tilde{K}_{P,o}(x,x) = \sum_{\gamma M_P(\mathbb{Q})\cap o} \sum_{\eta\in M_P(\mathbb{Q})} f(x^{-1}\eta^{-1}\gamma\eta x),$$

which is obtained when o consists entirely of semi-simple elements, can be understood better once we know the structure of $M_P(\mathbb{Q})\cap o$.

For any arbitrary parabolic subgroup P, let $W(P_1,P)$ be the set of elements $s\in S_k$ (the symmetric group of k elements) such that if $(r_{s(1)},\dots,r_{s(k)})$ determines P_1, then

$P_1 \subset P$, and $s^{-1}\alpha$ is a positive root for every $\alpha \in \Delta^P_{P_1}$. This is equivalent to saying that the parabolic subgroup P corresponds to the partition $(r_{s(1)} + \ldots + r_{s(i_1)}, r_{s(i_1+1)} + \ldots + r_{s(i_2)}, \ldots)$ for some $i_1, i_2, \ldots$ and $s^{-1}(i) < s^{-1}(i+1)$ for any $1 \leq i \leq k$, $i \neq i_1, \ldots, i_r$.

Now, for any s, let ω_s be the corresponding permutation matrix in $G(\mathbb{Q})$.

Exercise (5.13). Show that any element $\gamma \in M_P(\mathbb{Q}) \cap o$ can be represented uniquely in the form $\mu\omega_s\gamma_1\omega_s^{-1}\mu^{-1}$, $s \in W(P_1, P)$, $\mu \in M_{P,\omega_s\gamma\omega_s^{-1}}(\mathbb{Q})\backslash M_P(\mathbb{Q})$, where γ_1 is the representative of the semi-simple conjugacy class.

By the exercise above we have

$$\begin{aligned}\tilde{K}_{P,o}(x,x) &= \sum_{\gamma \in M_P(\mathbb{Q}) \cap o} \sum_{\eta \in M_P(\mathbb{Q})} f(x^{-1}\eta^{-1}\gamma\eta x) \\ &= \sum_{s \in W(P_1,P)} \sum_{\mu \in M_{P,\omega_s\gamma_1\omega_s^{-1}}(\mathbb{Q})\backslash M_P(\mathbb{Q})} \sum_{\eta} f(x^{-1}\eta^{-1}\mu^{-1}\omega_s\gamma_1\omega_s^{-1}\mu\eta x) \\ &= \sum_{s \in W(P_1,P)} \sum_{\pi \in M_{P,\omega_s\gamma_1\omega_s^{-1}}(\mathbb{Q})\backslash P(\mathbb{Q})}\end{aligned}$$

Therefore

$$\begin{aligned}\tilde{k}^T_o(x,f) &= \sum_P (-1)^{\dim(A_P/A_G)} \sum_{\delta \in P(\mathbb{Q})\backslash G(\mathbb{Q})} \\ &\quad \tilde{K}_{P,o}(\delta x, \delta x)\hat{\tau}_P(H_P(\delta x) - T) \\ &= \sum_P (-1)^{\dim(A_P/A_G)} \sum_{s \in W(P_1,P)} \sum_{\delta \in M_{P_1,\omega_s\gamma_1\omega_s^{-1}}(\mathbb{Q})\backslash G(\mathbb{Q})} \\ &\quad f(x^{-1}\delta^{-1}\omega_s\gamma_1\omega_s^{-1}\delta x)\hat{\tau}_P(H_P(\delta x) - T).\end{aligned}$$

Since o consists of a single semi-simple conjugacy class in $G(\mathbb{Q})$, the element γ_1 above is an element in $M_{P_1}(\mathbb{Q}) \cap o$, which satisfies the property that the centralizer of γ_1 is contained in M_{P_1}.

This property will be singled out as a definition. We also observe that this property is basic for what follows. In fact, to have a clearer formula for $\tilde{k}^T_o(x,f)$, put $\delta_1 = \omega_s^{-1}\delta$ and observe that $G_{\gamma_1} = M_{P,\gamma_1}$, and

$$\omega_s^{-1}(M_{P,\omega_s\gamma_1\omega_s^{-1}}(\mathbb{Q}))\omega_s = G_{\gamma_1}(\mathbb{Q}).$$

Then

$$\begin{aligned}\tilde{k}^T_o(x,f) &= \sum_P (-1)^{\dim(A_P/A_G)} \sum_{x \in W(P_1,P)} \sum_{\delta \in G_{\gamma_1}(\mathbb{Q})\backslash G(\mathbb{Q})} \\ &\quad f(x^{-1}\delta_1^{-1}\gamma_1\delta_1 x)\hat{\tau}_P(H_P(\omega_s\delta_1 x) - T).\end{aligned}$$

Which we may even write

$$\tilde{k}^T_o(x,f) = \sum_{\delta_1 \in G_{\gamma_1}\backslash G(\mathbb{Q})} f(x^{-1}\delta_1^{-1}\gamma_1\delta_1 x)\chi_T(\delta_1 x),$$

where

$$\begin{aligned}\chi_T(y) &= \sum_P (-1)^{\dim(A_P/A_G)} \sum_{s\in W(P_1,P)} \hat{\tau}_P(H_P(\omega_s y) - T) \\ &= \sum_{s\in S_k} \sum_{\{P: s\in W(P_1;P)\}} (-1)^{\dim(A_P/A_G)} \hat{\tau}_P(H_{P_o}(\omega_s y) - T).\end{aligned}$$

From this we get

$$\begin{aligned}J_o^T(f) &= \int_{G(\mathbb{Q})\backslash G(\mathbf{A})^1} \tilde{k}_o^T(x,f)dx \\ &= \int_{G(\mathbb{Q})\backslash G(\mathbf{A})^1} f(x^{-1}\gamma_1 x)\chi_T(x)dx. \end{aligned} \tag{5.3}$$

This is an expression for $J_o^T(f)$. However, we derive another expression for the same distribution in (5.4) below.

The fact that $G_{\gamma_1} \subseteq M_{P_1}$, implies $G_\gamma(\mathbf{A}) = G_{\gamma_1}(\mathbf{A})^1 \times A_{P_1}(\mathbb{R})^o$, where, $G_{\gamma_1}(\mathbf{A})^1 = G_\gamma(\mathbf{A}) \cap M_{P_1}(\mathbf{A})^1$. Hence x can be written as $x = may$, $m \in G_{\gamma_1}(\mathbb{Q})\backslash G_\gamma(\mathbf{A})^1$, $a \in A_{P_1}(\mathbb{R})^o \cap G(\mathbb{R})^1$, $y \in G_{\gamma_1}(\mathbf{A}) \cap G(\mathbf{A})^1 \backslash G(\mathbf{A})^1 \cong G_{\gamma_1}(\mathbf{A})\backslash G(\mathbf{A})$. Set $m' = \omega_s m \omega_s^{-1}$, then $m' \in \omega_s G_{\gamma_1}(\mathbf{A})^1 \omega_s^{-1} \subseteq M_P(\mathbf{A})^1$ as $s \in W(P_1, P)$. From the equality $\hat{\tau}_P(H_P(\omega_s may) - T)$ $= \hat{\tau}_P(H_P(\omega_s m \omega_s^{-1} \omega_s ay) - T)$ it follows that $\chi_T(x) = \chi_T(ay)$, hence

$$J_o^T(f) = \mathrm{vol}(G_{\gamma_1}(\mathbb{Q})\backslash G_{\gamma_1}(\mathbf{A})^1) \int_{G_{\gamma_1}(\mathbf{A})\backslash G(\mathbf{A})^1} f(y^{-1}\gamma_1 y) v_T(y)dy, \tag{5.4}$$

where, $\mathrm{vol}(G_{\gamma_1}(\mathbb{Q})\backslash G_{\gamma_1}(\mathbf{A})^1) < \infty$, and

$$v_T(y) = \int_{A_{P_1}(\mathbb{R})^o \cap G(\mathbb{R})^1} \chi_T(ay)da.$$

We remark that the groups $A_{P_1}(\mathbb{R})^o \cap G(\mathbb{R})^1$ and $\mathfrak{a}_{P_1}^G$ are isomorphic where the isomorphism map is H_{P_1}, and that

$$\begin{aligned}H_{P_o}(\omega_s ay) - T &= H_{P_o}(\omega_s a \omega_s^{-1} \omega_s y) - T \\ &= H_{P_o}(\omega_s a \omega_s^{-1}) + H_{P_o}(\omega_s y) - T \\ &= H_{P_o}(sH_{P_1}(a)) + H_{P_o}(\omega_s y) - T\end{aligned}$$

where s acts by permutation on $\mathfrak{a}_{P_1} \subseteq \mathfrak{a}_{P_o}$. Therefore

$$v_T(y) = \int_{\mathfrak{a}_{P_1}^G} \left[\sum_{s\in S_k} \sum_{\{P: s\in W(P_1,P)\}} (-1)^{\dim(A_P/A_G)} \hat{\tau}_P(sH + H_{P_o}(\omega_s y) - T) \right] dH.$$

To obtain a better understanding of the expression in the square brackets, recall that

$$W(P_1, P) = \{s \in S_k : \omega_s M_{P_1} \omega_s^{-1} = M_{P_1} \subseteq N_P;\ \ s^{-1}\alpha > 0\ \forall \alpha \in \Delta_{P_1'}^P\}.$$

Also there is a one-to-one correspondence between $\{Q : Q \supset P_1'\}$ and all subsets of $\Delta_{P_1'}$. Given $s \in S_k$, define a parabolic subgroup $P_s \supset P_1'$ by $\Delta_{P_1'}^{P_s} = \{\alpha \in \Delta_{P_1'} : s^{-1}\alpha > 0\}$. Then P ranges over all groups $P_1 \subset P \subset P_s$, which corresponds bijectively to the subsets of $\Delta_{P_1'}^{P_s}$. Therefore in the expression inside the brackets, the sum over P is just the sum over $\{P : P_1' \subset P \subset P_s\}$, which is an alternating sum of characteristic functions, hence by the alternating summation formula (see Exercise (3.9)) the sum over P vanishes for a given s, unless precisely one summand is non-zero.

On the other hand, the definition of $\hat{\tau}_P$ can be written in the following form. First, let χ' be the characteristic function of $\{H' : \tilde{\omega}(H') > 0 \ \forall \tilde{\omega} \in \hat{\Delta}_P - \hat{\Delta}_{P_s}\}$. Then

$$\hat{\tau}_P = \hat{\tau}_{P_s} \times \chi'.$$

Hence, we observe that the expression in the brackets is nothing but

$$(-1)^{\dim(A_{P_s}/A_G)} \hat{\tau}_s(sH + H_{P_0}(\omega_s y) - T),$$

where $\hat{\tau}_s$ is the characteristic function of

$$\{X \in \mathbf{a}_{P_0} : \tilde{\omega}(X) > 0 \ \forall \tilde{\omega} \in \hat{\Delta}_{P_s}, \ \tilde{\omega}(X) \leq 0 \ \forall \tilde{\omega} \in \hat{\Delta}_{P_0} - \hat{\Delta}_{P_0}\}.$$

Now, we need the following result, which is a consequence of the combinatorial lemma of Langlands [68], and it is also proven in [5].

Lemma (5.14). The function

$$\sum_{s \in S_k} (-1)^{\dim(A_{P_s}/A_G)} \hat{\tau}_s(sH + H_{P_0}(\omega_s y) - T),$$

where $H \in \mathbf{a}^G_{P_1}$, is the characteristic function of the projection onto $\mathbf{a}^G_{P_1}$ of the convex hull of $\{s^{-1} H_{P_0}(\omega_s y) - s^{-1} T = Y_s, \ s \in S_k\}$.

Example (5.15). Let $r = 3$, $G = GL(3)$, $P_1 = P_0$, $\dim \mathbf{a}^G_{P_1} = 2$, $\Delta_{P_0} = \{\alpha_1, \alpha_2\}$, then $\hat{\tau}_s(sH + H_{P_0}(\omega_s y) - T) =$ characteristic function of

$$\{-Y_s + t_1 s^{-1}(\alpha_1) + t_2 s^{-1}(\alpha_2)\}$$

such that

$$t_i < 0 \text{ if } s^{-1}(\alpha_i) > 0, \ t_i > 0 \text{ if } s^{-1}(\alpha_i) < 0, \ (i = 1, 2).$$

Now Lemma (5.14) implies that $v_T(y)$ is the volume in $\mathbf{a}^G_{P_1}$ of the projection onto $\mathbf{a}^G_{P_2}$ of the convex hull of $\{Y_s : s \in S_k\}$, which is a polynomial in T, since it is given by the integral of a characteristic function in variable T. Therefore, we get the following expression for $J_o(f)$.

$$J_0(f) = \mathrm{vol}(G_{\gamma_1}(\mathbf{Q}) \backslash G_{\gamma_1}(\mathbf{A})^1) \int_{G_{\gamma_1}(\mathbf{A}) \backslash G(\mathbf{A})} f(y^{-1} \gamma_1 y) v(y) dy. \tag{5.5}$$

For our future purposes, it is sometimes more convenient to take the set of parabolic subgroups defined in the following manner.

First we have $W_0 = S_k$, then $s \in S_k$, gives rise to a permutation matrix ω_s. Let $\mathbf{F} = \{\omega_s P \omega_s^{-1} : P \text{ as before (i.e. } P \supseteq P_0, s \in S_k)\}$, so that $\mathbf{F}$ is the set of parabolic subgroups of G which contain the groups of diagonal matrices (sometimes called *semistandard* parabolic subgroups, and the parabolic subgroups that contain P_0 are called *standard* parabolic subgroups). Moreover, let $\mathbf{L} = \{\omega_s M_P \omega_s^{-1} : P \text{ as before, and } s \in S_r\}$. Then if

$Q \in \mathbf{F}$, we have $Q = M_Q N_Q$; where $M_Q \in \mathbf{L}$. If $M \in \mathbf{L}$, write $\mathbf{F}(M) = \{Q \in \mathbf{F} : Q \supset M\}$ $= \{Q \in \mathbf{F} : M_Q \supset M\}$. Write $\mathbf{P}(M) = \{Q \in \mathbf{F} : M_Q = M\}$, and $A_M = A_Q$, $\mathbf{a}_M = \mathbf{a}_Q$ (for any $Q \in \mathbf{P}(M)$). Suppose that P_1 corresponds to the partition $(r_1, \ldots, r_k)$, $s \in S_k$ and P corresponds to the partition $(r_{s(1)}, \ldots, r_{s(k)})$. Let $M = M_P$. Then $\omega_s^{-1} P_1 \omega_s = P$ belongs to $\mathbf{P}(M)$, and any element of $\mathbf{P}(M)$ arises in this way.

Exercise (5.15). Prove that for $GL(r)$, $s^{-1} H_{P_1'}(\omega_s x) = H_P(x)$, where for $x = nmk \in G(\mathbf{A})$ one define $H_P(x) = H_M(m)$.

By the exercise above we obtain the following *weight function*

$$\begin{aligned} v(x) &= v_M(x) \\ &= \text{volume of the convex hull of } \{H_P(x) : P \in \mathbf{P}(M)\}. \end{aligned} \tag{5.6}$$

Hence we have proved that if $o = \{\gamma_1\}$, then

$$J_o(f) = \mathrm{vol}(G_{\gamma_1} \backslash G_{\gamma_1}(\mathbf{A})^1) \int_{G_{\gamma_1}(\mathbf{Q}) \backslash G(\mathbf{A})^1} f(x^{-1}\gamma_1 x) v_M(x) dx. \tag{5.7}$$

As another example, in the case of $GL(3)$, see the figure below (cf. [6]). Note that the Weyl group in this case is S_3.

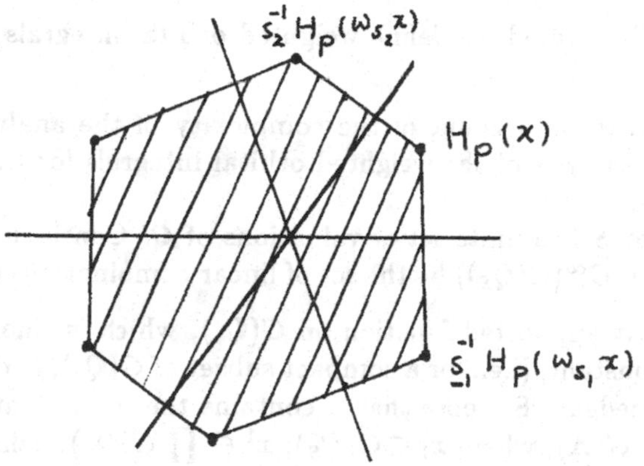

To obtain an expression for $J_o(f)$ when o is a general arbitrary class, one may first consider the extreme case, i.e., when o consists entirely of unipotent elements in $G(\mathbf{Q})$, i.e., $o = \{\gamma \in G(\mathbf{Q}) : \gamma_s = 1\}$. This study then leads to the local investigations of weighted orbital integrals.

Chapter VI

The Geometric Expansion Of The Trace Formula

We have already seen that Selberg's trace formula is an identity which consists of two expansions. One of it was called the geometric expansion, and the other, the spectral expansion. In the case of Arthur's trace formula, the situation is similar, i.e., we have also an identity between two expressions, one is the geometric expansion based on the distribution $J_o^T(f)$, and the other is the spectral expansion based on the analysis of the Eisenstein series. The purpose of this chapter is to explain the geometric expansion of Arthur's trace formula. This, as we shall see, is based on the local study of weighted orbital integrals, predicted by Langlands. Weighted orbital integrals were presented through the introduction of the weight function (see (5.5), (5.7)). Those orbital integrals were the expression of $J_o^T(f)$, for the semi-simple conjugacy classes o. In the first section of this chapter, we define weighted orbital integrals for the unipotent classes.

6.1. Weighted orbital integrals

We begin this section with the following question.

Question 1. How could we define weighted orbital integrals for a unipotent element γ?

To answer this, on account of the complexity of the analysis involved, one has to exploit the full analysis of the weighted orbital integrals for the local fields $\mathbb{R}$, and $\mathbb{Q}_P$ with $p < \infty$, [8].

Suppose that S is a finite set of valuations of $\mathbb{Q}$. Consider the locally compact ring $\mathbb{Q}_S = \prod_{v \in S} \mathbb{Q}_v$. Let $C_c^\infty(G(\mathbb{Q}_S))$ be the set of linear combinations of functions $\bigotimes_{v \in S} f_v$, where f_v is a compactly supported function on $G(\mathbb{Q}_v)$, which is smooth if v is Archimedean, and is locally constant, (i.e., for a compact subset of $G(\mathbb{Q}_v)$, f_v on that set is constant), if v is non-Archimedean. Suppose that S contains the real valuations. If $f \in C_c^\infty(G(\mathbb{Q}_v))$, and $x = x_1 x^1 \in G(\mathbf{A})$, where $x_1 \in G(\mathbb{Q}_S)$, $x^1 \in \prod_{v \notin S} G(\mathbb{Q}_S)$, define

$$f(x) = \begin{cases} f(x_1) & \text{if} \quad x^1 \in \prod_{v \notin S} K_v, \\ 0 & \text{if} \quad x^1 \notin \prod_{v \notin S} K_v. \end{cases}$$

In this case, f is a function on $C_c^\infty(G(\mathbf{A}))$, where

$$C_c^\infty(G(\mathbf{A})) = \varinjlim_{S} C_c^\infty(G(\mathbb{Q}_S)).$$

Now we want to define $J_o(f)$ indirectly. For this we obtain an expression by means of the new distributions $J_M(\gamma, f)$. These are defined as follows:

Let $\gamma = \prod_{v\in S} \gamma_v \in G(\mathbb{Q}_v)$; write $G_\gamma(\mathbb{Q}_S) = \prod_{v\in S} G_{\gamma_v}(\mathbb{Q}_v)$. Let $\mathcal{G}_v(\mathbb{Q}_v)$ be the Lie algebra of $G_{\gamma_v}(\mathbb{Q}_v)$, the centralizer of γ_v in the Lie algebra $M_r(\mathbb{Q}_v)$. If $\gamma_v = \sigma_v\, u_v$ is the Jordan decomposition, (note the change of symbols), and $M \in \mathbf{L}$, $\gamma \in M(\mathbb{Q}_S)$, $f \in C_c^\infty(G(\mathbb{Q}_S))$, define

$$J_M(\gamma, f) = |\det(1 - Ad(\gamma))_{\mathcal{G}/\mathcal{G}_\gamma}|^{\frac{1}{2}} \int_{G_\gamma(\mathbb{Q}_S)\backslash G(\mathbb{Q}_S)} f(x^{-1}\gamma x) v_M(x)dx, \qquad (6.1)$$

provided that $G_\gamma(\mathbb{Q}_S) \subset M(\mathbb{Q}_S)$ and

$$|\det(1 - Ad(\gamma))_{\mathcal{G}/\mathcal{G}_\gamma}|^{\frac{1}{2}} = \prod_{v\in S} |\det(1 - Ad(\sigma_v))_{M_r(\mathbb{Q}_v)/\mathcal{G}_{\sigma_v}}|_v^{\frac{1}{2}}.$$

We note that if $m \in M(\mathbb{Q}_S)$, $v_M(mx)$ equals the volume of the convex hull $\{H_P(mx) : P \in \mathbf{P}(M)\}$ which in turn equals the volume of the convex hull $\{H_M(m) + H_P(x) : P \in \mathbf{P}(M)\}$, but $H_M(m)$ represents a translation of the convex hull, therefore $v_M(mx) = v_M(x)$. For the purposes of our study, it remains for us to ask:

Question 2. (i) Can one define $J_M(\gamma, f)$ even if, $G_\gamma(\mathbb{Q}_S) \not\subset M(\mathbb{Q}_S)$?

(ii) Can one express $J_O(f)$, in terms of $J_M(\gamma, f)$?

It turns out that the answers to these questions are positive. To explain this, write $P = MN_P \in \mathbf{P}(M)$, with $A_M = A_P$ (center of M). Suppose that γ is an element in $M(\mathbb{Q}_S)$, and $a \in A_M(\mathbb{Q}_S) = \prod_{v\in S} A_M(\mathbb{Q}_v)$ which is close to one but in *general position* i.e., $G_{a\gamma} = M_\gamma$. Then $G_{a\gamma}(\mathbb{Q}_S) = M_\gamma(\mathbb{Q}_S)$, since γ and a commute. Therefore $J_M(a\gamma, f)$ is well defined. So, under our condition on a, we may now ask. Does $\lim_{a\to 1} J_M(a\gamma, f)$ existe ?

The answer will be provided by some general remarks and an example. First, write $M \cong GL(r_1) \times \ldots \times GL(r_k)$, $M \in \mathbf{L}$. Then

(i) Set $W_0^M = S_{r_1} \times S_{r_2} \times \ldots \times S_{r_k}$ (Weyl group).

(ii) For any $o \in O$, we have

$$J_o(f^y) = \sum_{Q\in\mathbf{F}} |W_0^{M_Q}|\,|W_0^G|^{-1} J_o^{M_Q}(f_Q, y), \quad y \in G(\mathbb{Q}_S)$$

(compare with formula (5.1) where $J_o^{M_Q} = J_{o\cap M_Q}^{M_Q}$).

(iii) For $M \in \mathbf{L}, \gamma \in M(\mathbb{Q}_S), G_\gamma \subset M$, we have

$$J_M(\gamma, f^y) = \sum_{Q\in\mathbf{F}(M)} J_M^{M_Q}(\gamma, f_{Q,y}),$$

where $J_M^{M_Q}$ is the same distribution, but on $M_Q(\mathbb{Q}_S)$ instead of $G(\mathbb{Q}_S)$ evaluated at function $f_{Q,y} \in C_c^\infty(M_Q(\mathbb{Q}_S))$, (compare with formula (6.1)). One may also see the proof in [4], section 8.

Example (6.1). Let $G = GL(2)$, and suppose that S contains only Archimedean valuations. Then $\mathbb{Q}_S = \mathbb{R}$, $M = M_{P_0}$, and for γ take $\gamma = \begin{pmatrix} a & 0 \\ 0 & b \end{pmatrix}$, $a, b \in \mathbb{R}^*$

with $a \neq b$. According to our assumptions M consists of the diagonal elements and $G_\gamma(\mathbb{R}) = M(\mathbb{R}) = A_{P_0}(\mathbb{R})$, moreover $\mathbf{P}(M)$ consists of upper triangular matrices P_0 and lower triangular matrices $\bar{P}_0$ of $GL(2)$. Then

$$\begin{aligned} G_\gamma(\mathbb{R})\backslash G(\mathbb{R}) &= A_{P_0}(\mathbb{R})\backslash A_{P_0}(\mathbb{R})N_{P_0}(\mathbb{R})K_{\mathbb{R}} \\ &\cong N_{P_0}(\mathbb{R})K_{\mathbb{R}}. \end{aligned}$$

For this example, it is very easy to compute $v_M(x)$ when $x \in G_\gamma(\mathbb{R})\backslash G(\mathbb{R})$ i.e., $x = nk$, $n \in N_{P_0}(\mathbb{R})$, $k \in K_{\mathbb{R}}$. $v_M(x)$ is then the volume of the convex hull of $\{H_{P_0}(x), H_{\bar{P}_0}(x)\}$ in $\mathfrak{a}^G_{M_0} \cong \mathbb{R}$. But $H_{P_0}(x) = 0$, therefore $\{H_{P_0}(x), H_{\bar{P}_0}(x)\} = \{0, H_{\bar{P}_0}(x)\}$, hence

$$v_M(x) = \| H_{\bar{P}_0}(x) \| .$$

To compute $H_{\bar{P}_0}(x)$, we recall that $x = \begin{pmatrix} 1 & t \\ 0 & 1 \end{pmatrix}$ with $t \in \mathbb{R}$, which may be written as

$$x = \begin{pmatrix} 1 & t \\ 0 & 1 \end{pmatrix} = \begin{pmatrix} 1 & 0 \\ y & 1 \end{pmatrix}\begin{pmatrix} a & 0 \\ 0 & b \end{pmatrix}k$$

with $k \in O(2)$, i.e., $k = \begin{pmatrix} \alpha & -\beta \\ \beta & \alpha \end{pmatrix}$ with $\alpha^2 + \beta^2 = 1$.

So that

$$\begin{aligned} x &= \begin{pmatrix} a & 0 \\ ya & b \end{pmatrix}\begin{pmatrix} \alpha & -\beta \\ \beta & \alpha \end{pmatrix} \\ &= \begin{pmatrix} a\alpha & -a\beta \\ ya\alpha + b\beta & -y\beta a + b\alpha \end{pmatrix}. \end{aligned}$$

This is a system of equations which gives

$$a = \sqrt{1+t^2}, \quad b = \frac{1}{a} = \frac{1}{\sqrt{1+t^2}}.$$

Therefore

$$H_{\bar{P}_0}(x) = H_{\bar{P}_0}\begin{pmatrix} 1 & t \\ 0 & 1 \end{pmatrix} = \log\sqrt{1+t^2}.$$

This formula can also be verified by showing that

$$H_{\bar{P}_0}\begin{pmatrix} 1 & t \\ 0 & 1 \end{pmatrix} = -\log \left\| \begin{pmatrix} 1 & t \\ 0 & 1 \end{pmatrix}^{-1}\begin{pmatrix} 0 \\ 1 \end{pmatrix} \right\| .$$

To see this, we must note that

$$\begin{aligned} \left\| \begin{pmatrix} 1 & t \\ 0 & 1 \end{pmatrix}^{-1}\begin{pmatrix} 0 \\ 1 \end{pmatrix} \right\| &= \left\| k^{-1}\begin{pmatrix} a^{-1} & 0 \\ 0 & b^{-1} \end{pmatrix}\begin{pmatrix} 1 & 0 \\ -y & 1 \end{pmatrix}\begin{pmatrix} 0 \\ 1 \end{pmatrix} \right\| \\ &= \left\| k^{-1}\begin{pmatrix} a^{-1} & 0 \\ -yb^{-1} & b^{-1} \end{pmatrix}\begin{pmatrix} 0 \\ 1 \end{pmatrix} \right\| \\ &= |b|^{-1}. \end{aligned}$$

Hence,

$$H_{P_0}\begin{pmatrix} 1 & t \\ 0 & 1 \end{pmatrix} = -\log|b|^{-1}$$

$$= \log|b|.$$

However, the projection of $\begin{pmatrix} \log a & 0 \\ 0 & \log b \end{pmatrix}$ onto a_M^G is $-\log|b|$. Therefore

$$H_{P_0}\left(\begin{pmatrix} 1 & t \\ 0 & 1 \end{pmatrix}\right) = \log \left\| \begin{pmatrix} 1 & -t \\ 0 & 1 \end{pmatrix}\begin{pmatrix} 0 \\ 1 \end{pmatrix} \right\| = \log \left\| \begin{pmatrix} 1 \\ -t \end{pmatrix} \right\| = \log\sqrt{1+t^2}.$$

Hence,

$$v_M(x) = v_M\begin{pmatrix} 1 & t \\ 0 & 1 \end{pmatrix} = \log\sqrt{1+t^2}.$$

So, to obtain a formula for $J_M(\gamma, f)$ we need to calculate $\det|(1 - Ad(\gamma)_{\mathcal{G}/g_\gamma}|^{\frac{1}{2}}$. For γ, as above, the Lie algebra $\mathcal{G}/g_\gamma$ is identified with the Lie algebra of $\begin{pmatrix} 0 & * \\ * & 0 \end{pmatrix}$ type which has as its basis the matrices $\begin{pmatrix} 0 & 1 \\ 0 & 0 \end{pmatrix}$, and $\begin{pmatrix} 0 & 0 \\ 1 & 0 \end{pmatrix}$. $Ad(\gamma)$ then acts on $\mathcal{G}/g_\gamma$ by the conjugation of $\gamma = \begin{pmatrix} a & 0 \\ 0 & b \end{pmatrix}$ on the basis elements i.e.,

$$\begin{pmatrix} a & 0 \\ 0 & b \end{pmatrix}\begin{pmatrix} 0 & 1 \\ 0 & 0 \end{pmatrix}\begin{pmatrix} a^{-1} & 0 \\ 0 & b^{-1} \end{pmatrix} = \frac{a}{b}\begin{pmatrix} 0 & 1 \\ 0 & 0 \end{pmatrix},$$

and

$$\begin{pmatrix} a & 0 \\ 0 & b \end{pmatrix}\begin{pmatrix} 0 & 0 \\ 1 & 0 \end{pmatrix}\begin{pmatrix} a^{-1} & 0 \\ 0 & b^{-1} \end{pmatrix} = \frac{b}{a}\begin{pmatrix} 0 & 0 \\ 1 & 0 \end{pmatrix}.$$

Therefore

$$Ad(\gamma) = \begin{pmatrix} ab^{-1} & 0 \\ 0 & ba^{-1} \end{pmatrix},$$

so that

$$|\det(I - Ad(\gamma))|^{\frac{1}{2}} = |(1 - ab^{-1})(1 - ba^{-1})|^{\frac{1}{2}}$$

$$= \left(\frac{|a-b|^2}{|ab|}\right)^{\frac{1}{2}} = \frac{|a-b|}{\sqrt{|ab|}}.$$

This implies

$$J_M(\gamma, f) = \frac{|a-b|}{\sqrt{|ab|}}\int_{O(2,\mathbb{R})}\int_{\mathbb{R}}$$

$$f\left(k^{-1}\begin{pmatrix} 1 & -t \\ 0 & 1 \end{pmatrix}\begin{pmatrix} a & 0 \\ 0 & b \end{pmatrix}\begin{pmatrix} 1 & t \\ 0 & 1 \end{pmatrix}k\right)\log\sqrt{1+t^2}\,dt\,dk.$$

We may now look at the possibility of obtaining a formula for $J_M(\gamma, f)$ when $\gamma \to 1$. This can be done by letting $a \to b$ in γ. To begin with,

$$\begin{pmatrix} 1 & -t \\ 0 & 1 \end{pmatrix}\begin{pmatrix} a & 0 \\ 0 & b \end{pmatrix}\begin{pmatrix} 1 & t \\ 0 & 1 \end{pmatrix} = \begin{pmatrix} a & 0 \\ 0 & b \end{pmatrix}\begin{pmatrix} 1 & t(1 - a^{-1}b) \\ 0 & 1 \end{pmatrix}.$$

Let $t_1 = t(1 - a^{-1}b)$. Then $dt_1 = dt(1 - a^{-1}b)$ and

$$\begin{aligned} J_M(\gamma, f) &= |ab^{-1}|^{\frac{1}{2}} \int_K \int_{\mathbb{R}} \\ & f\left(k^{-1}\begin{pmatrix} a & 0 \\ 0 & b \end{pmatrix}\begin{pmatrix} 1 & t_1 \\ 0 & 1 \end{pmatrix} k\right) \log\sqrt{1 + \left(\frac{t_1}{1 - a^{-1}b}\right)^2}\, dt_1 dk \\ &= |ab^{-1}|^{\frac{1}{2}} \int_K \int_{\mathbb{R}} \\ & f\left(k^{-1}\begin{pmatrix} a & 0 \\ 0 & b \end{pmatrix}\begin{pmatrix} 1 & t_1 \\ 0 & 1 \end{pmatrix} k\right) \Big[\log\sqrt{(1 - a^{-1}b)^2 + t_1^2} \\ & \log(|1 - a^{-1}b|) \Big]\, dt_1 dk. \end{aligned}$$

Clearly this expression does not have a limit as $a \to b$. The only term that causes trouble is

$$\begin{aligned} & |ab^{-1}|^{\frac{1}{2}} \int_K \int_{\mathbb{R}} f\left(k^{-1}\begin{pmatrix} a & 0 \\ 0 & b \end{pmatrix}\begin{pmatrix} 1 & t_1 \\ 0 & 1 \end{pmatrix} k\right) dt_1 dk(\log(|1 - a^{-1}b|)) \\ &= \left(\frac{|a-b|}{|ab|}\right)^{\frac{1}{2}} \int_K \int_{\mathbb{R}} f\left(k^{-1}\begin{pmatrix} 1 & -t \\ 0 & 1 \end{pmatrix}\begin{pmatrix} a & 0 \\ 0 & b \end{pmatrix}\begin{pmatrix} 1 & t \\ 0 & 1 \end{pmatrix} k\right) \times \\ & \qquad dt dk(\log|1 - a^{-1}b|) \\ &= \log(|1 - a^{-1}b|) J_G(\gamma, f), \end{aligned}$$

where $J_G(\gamma, f)$ is an (unweighted) orbital integral. Thus

$$\begin{aligned} & |a^{-1}b|^{\frac{1}{2}} \int_K \int_{\mathbb{R}} f\left(k^{-1}\begin{pmatrix} a & 0 \\ 0 & b \end{pmatrix}\begin{pmatrix} 1 & t_1 \\ 0 & 1 \end{pmatrix} k\right) \log\sqrt{(1 - a^{-1}b)^2 + t_1^2}\, dt_1 dk \\ &= J_M(\gamma, f) + \log(|1 - a^{-1}b|) J_G(\gamma, f). \end{aligned}$$

Now set $\gamma_0 = \begin{pmatrix} a & 0 \\ 0 & a \end{pmatrix}$, then as $a \to b$, $\gamma \to \gamma_0$, the above limit exists and it is in fact equal to

$$\int_K \int_{\mathbb{R}} f\left(k^{-1}\begin{pmatrix} b & 0 \\ 0 & b \end{pmatrix}\begin{pmatrix} 1 & t_1 \\ 0 & 1 \end{pmatrix} k\right) \log|t_1| dt_1 dt \stackrel{\text{def}}{=} J_M(\gamma_0, f).$$

This allows us to write

$$J_M(\gamma_0, f) = \lim_{\gamma \to \gamma_0} (J_M(\gamma, f) + \log(|1 - a^{-1}b|) J_G(\gamma, f)).$$

This result can be better explained and generalized in the following Theorem.

Theorem (6.2). Let S, M be as before, $\gamma \in M(\mathbb{Q}_S)$. Then there are functions $r_M^L(\gamma, a)$,

$L \in \mathbf{L}(M)$, where a is an element in $A_M(\mathbf{Q}_S)$ close to 1, but in general position (i.e., $G_{a\gamma} = M_\gamma$) such that

$$\lim_{a \to 1} \sum_{L \in \mathbf{L}(M)} r_M^L(\gamma, a) J_L(a\gamma, f) \stackrel{\text{def}}{=} J_M(\gamma, f)$$

whenever the limit exists.

Example (6.3). In the case of $GL(2)$, $r_M^L(\gamma, a)$ is identity. And in the case of $GL(3)$, these functions were first obtained by Flicker [40].

Remark. (1) The proof of the theorem above can be found in [8].

(2) For a as in the theorem, $G_{a\gamma}(\mathbf{Q}_S) = M_\gamma(\mathbf{Q}_S) \subseteq M(\mathbf{Q}_S)$, so that $J_M(a\gamma, f)$ is defined as before. $J_M(\gamma, f)$ is a weighted orbital integral for general γ. We recall that it was already conjectured by Langlands that the terms in a general trace formula must be given in the form of weighted orbital integrals (see [1]).

(3) If $G_\gamma \subseteq M$, $J_M(\gamma, f^v) = \sum_{Q \in \mathbf{F}(M)} J_M^{M_Q}(\gamma, f_{Q,v})$.

Here, it is natural to ask about the validity of formula (3) in the remark above when $\gamma \in M(\mathbf{Q}_S)$ is arbitrary.

Let us first consider the case $G_{a\gamma} \subseteq M \subseteq L$. In this case

$$\begin{aligned} J_M(\gamma, f^v) &= \lim_{a \to 1} \sum_{L \in \mathbf{L}(M)} r_M^L(\gamma, a) J_L(a\gamma, f^v) \\ &= \lim_{a \to 1} \left(\sum_{L \supset M} \sum_{Q \supset L} r_M^L(\gamma, a) J_L^{M_Q}(a\gamma, f_{Q,v}) \right) \\ &= \lim_{a \to 1} \left(\sum_{Q \supset M} \sum_{\{L : M \subset L \subset M_Q\}} r_M^L(\gamma, a) J_L^{M_Q}(a\gamma, f_{Q,v}) \right) \\ &= \sum_{Q \in \mathbf{F}(M)} \lim_{a \to 1} \left(\sum_{\{L : M \subset L \subset M_Q\}} r_M^L(\gamma, a) J_L^{M_Q}(a\gamma, f_{Q,v}) \right). \end{aligned}$$

But, $\{L : M \subset L \subset M_Q\} = \mathbf{L}^{M_Q}(M)$ (i.e., the set $\mathbf{L}(M)$, where G is replaced by M_Q). Hence, the summation above is by definition

$$\sum_{Q \in \mathbf{F}(M)} J_M^{M_Q}(\gamma, f_{Q,v}).$$

Therefore our formula in part (3) of the remark holds for any $\gamma \in M(\mathbf{Q}_S)$. That is to say, there is no need for the condition $M \supseteq G_\gamma$, and this provides the positive answer to the first part of the preceding question 2.

6.2. The unipotent distribution

This section is an immediate continuation of section 6.1. In fact, we want to understand the second part of question 2 of the preceding section. We first modify the question by replacing it with the following one. First, let $\{U_M(\mathbf{Q})\}$ be the finite set of conjugacy classes of unipotent matrices in $M(\mathbf{Q})$ diagnonally embedded in $M(\mathbf{Q}_S)$. It is clear that $\{U_M(\mathbf{Q})\}$ can be considered as rational points of an algebraic variety denoted by U_M. Arthur in his paper [7] studies certain measure and constants related to the following question.

Question 3. Let as before, S be a finite set of valuations of $\mathbf{Q}$ containing the Archimedean valuation. Let $f \in C_c^\infty(G(\mathbf{Q}_S)) \subseteq C_c^\infty(G(\mathbf{A}))$. Is $J_o(f) = J_{unip}(f)$ a linear combination of weighted orbital integrals $J_M(\gamma, f)$, with γ being a unipotent element in $M(\mathbf{Q}_S)$? In another words, can we find constants

$$\{a^M(S,u) : M \in \mathbf{L}, \quad u \in U_M(\mathbf{Q})\}$$

such that

$$J_o(f) = \sum_{M\in\mathbf{L}} |W_0^M||W_0^G|^{-1} \sum_{u\in\{U_M(\mathbf{Q})\}} a^M(S,u) J_M(u,f).$$

This question can actually be answered as follows.

Assume inductively that this is true if G is replaced by any $M \subsetneq G$. Then $a^M(S,u)$ is defined if $M \subsetneq G$.

Define

$$T(f) = J_o(f) - \sum_{\{M\in\mathbf{L}:M\neq G\}} |W_0^M||W_0^G|^{-1} \sum_{u\in\{U_M(\mathbf{Q})\}} a^M(S,u) J_M(u,f),$$

where $f \in C_c^\infty(G(\mathbf{Q}_S))$. We need to show that there are uniquely determined numbers $\{a^M(S,u)\}$ such that

$$T(f) = \sum_{u\in\{U_G(\mathbf{Q})\}} a^G(S,u) J_G(u,f), \tag{6.2}$$

for every f. For this, we compute the difference

$$T(f^y) - T(f).$$

This difference is given by

$$J_o(f^y) - J_o(f) - \sum_{M\neq G} |W_0^M||W_0^G|^{-1} \sum_u a^M(S,u)(J_M(u,f^y) - J_M(u,f))$$

$$= \sum_{\{Q\in\mathbf{F}:Q\neq G\}} |W_0^{M_Q}||W_0^G|^{-1} J_0^{M_Q}(f_{Q,y}) - \sum_{M\in\mathbf{L}} \sum_{\{Q\in\mathbf{F}(M):Q\neq G\}}$$

$$|W_0^M||W_0^G|^{-1} \sum_{u\in\{U_M(\mathbf{Q})\}} a^M(S,u) J_M^{M_Q}(u, f_{Q,y})$$

$$= \sum_{\{Q\in\mathbf{F}:Q\neq G\}} |W_0^{M_Q}||W_0^G|^{-1} \Bigg\{ J_o^{M_Q}(f_{Q,y}) - \sum_{\{M\in\mathbf{L}:M\subset M_Q\}} |W_0^M||W_0^{M_Q}|^{-1}$$

$$\sum_{u\in\{U_M(\mathbf{Q})\}} a^M(S,u) J_M^{M_Q}(u, f_Q) \Bigg\}.$$

Note that $L^{M_Q} = \{M \in \mathbf{L} : M \subset M_Q\}$ is the set $\mathbf{L}$, but with G replaced by M_Q.

Now by the inductive argument, the expression in brackets is zero. Therefore $T(f^y) = T(f)$. This means that T is an invariant distribution. Since $J_o(f)$ and $J_M(u,f)$ vanish if f vanishes on the unipotent set in $G(\mathbb{Q}_S)$, we must deduce that $T(f)$ vanishes too i.e., it annihilates any function f which vanishes on $U_G(\mathbb{Q}_S)$. For example, let us consider the special case of $S = \{|\cdot|_\infty\}$, $G = GL(2)$, then $\{U_G(\mathbb{Q})\} = \left\{\gamma_1 = \begin{pmatrix} 1 & 0 \\ 0 & 1 \end{pmatrix}, \gamma_2 = \begin{pmatrix} 1 & 1 \\ 0 & 1 \end{pmatrix}\right\}$. It is easy to check that

$$T(f) = c_1 f\begin{pmatrix} 1 & 0 \\ 0 & 1 \end{pmatrix} + c_2 \int_{G_{\gamma_2}(\mathbb{R})\backslash G(\mathbb{R})} f(x^{-1}\gamma_2 x)dx.$$

Similarly, in general, one can see that $T(f)$ may be written as in (6.2), where $a^G(S,u)$ are the same constants as before, and $J_G(u,f)$ are orbital integrals over u, i.e., $v_G(x) \equiv 1$, and $U_G(\mathbb{Q}_S)$ are unipotent classes in $GL(r, \mathbb{Q}_S)$. In fact, it is desired to have a formula for $T(f)$ as in (6.2) so that the sum in the formula runs over $u \in U_G(\mathbb{Q})$, the unipotent classes in $GL(r, \mathbb{Q})$, (cf. [7], page 1269).

Now we are in a position to state our theorem which gives a final form of $J_o(f)$ for general $o \in \mathcal{O}$.

Theorem (6.4). If S is a large finite set of valuations of $\mathbb{Q}$ which contains the Archimedean valuation, and $f \in C_c^\infty(G(\mathbb{Q}_S))$, then

$$J_o(f) = \sum_{M \in \mathbf{L}} |W_0^M||W_0^G|^{-1} \sum_{\gamma \in M(\mathbb{Q}) \cap o} a^M(S,\gamma) J_M(\gamma, f),$$

where $a^M(S,\gamma)$ are uniquely determined constants such that,

(i) if $\gamma = \sigma u$ is the Jordan decomposition, then

$$a^M(S,\gamma) = \begin{cases} a^M(S,u) & \text{if } \sigma \text{ is } \mathbb{Q}-\text{elliptic in } M, \\ 0 & \text{otherwise.} \end{cases}$$

(ii) if γ is semi-simple,

$$a^M(S,\gamma) = \begin{cases} \mathrm{vol}(M_\gamma(\mathbb{Q})\backslash M_\gamma(\mathbf{A})^1) & \text{if } \gamma \text{ is } \mathbb{Q}-\text{elliptic in } M, \\ 0 & \text{otherwise.} \end{cases}$$

Note that, in this theorem, an element $\gamma \in GL(r, \mathbb{Q})$ is called *$\mathbb{Q}-$elliptic* if $\gamma \in E^*$, where E is an extension field of $\mathbb{Q}$ of degree r, embedded in $GL(r, \mathbb{Q})$ in the usual way. See also ([11], Theorem 8.1).

Corollary (6.5). With the notations as above one has

$$J(f) = \sum_{M \in \mathbf{L}} |W_0^M||W_0^G|^{-1} \sum_{\gamma \in M(\mathbb{Q})} a^M(S,\gamma) J_M(\gamma, f).$$

Definition (6.6). The expression of $J(f)$ given in the corollary above is called the *geometric expansion of the trace formula.*

Remark (6.7). As Arthur shows in his paper ([7], Corollary 8.3), $J_o(f)$ when o is unipotent, defines a measure on $G(\mathbf{Q}_S)^1$. In fact, this measure is not invariant, but it is absolutely continuous with respect to the measure on the unipotent set invariant under $G(\mathbf{Q}_S)$. It is interesting to recall that using this measure, we can study the problem of limit of multiplicities of certain automorphic representations and Hecke operators ([31], [99]). See also [92] for a method which is not based on the trace formula.

Chapter VII

The Spectral Theory

The spectral theory to be discussed in this chapter is based on some results of Langlands on the Eisenstein series [68]. Eisenstein series are used to study the continuous spectrum of the regular representation, and as we shall see they completely characterize it. Moreover, by using the Eisenstein series, we can define certain kernel functions which are relevant (as we prefer to say "dual") to the kernel functions introduced so far. Thus, these kernel functions will be the main object of our study in order to derive an spectral expansion of Arthur's trace formula. These subjects will be discussed in two sections of this chapter, where the content of the second section is directly related to the spectral expansion of the trace formula, and thus completes the final expression of the non-invariant trace formula.

7.1. A review of the Eisenstein series

In this section and in the next one, we assume that R is the regular representation of $G(\mathbf{A})^1$ on the Hilbert space $L^2(G(\mathbb{Q})\backslash G(\mathbf{A})^1)$. The spectral decomposition of R consists of two parts, a discrete spectrum and a continuous spectrum. Hence, we can write

$$L^2(G(\mathbb{Q})\backslash G(\mathbf{A})^1) = L^2_{\text{disc}}(G(\mathbb{Q})\backslash G(\mathbf{A})^1) \oplus L^2_{\text{con}}(G(\mathbb{Q})\backslash G(\mathbf{A})^1),$$

where the first summand is the discrete spectrum, and the second, the continuous spectrum. The right hand side of the equality is called the *partial spectral decomposition.*

We mention that, if it is necessary we feel free to consider the Hilbert space

$$L^2(G(\mathbb{Q})A_G(\mathbb{R})^0\backslash G(\mathbf{A}))$$

in place of

$$L^2(G(\mathbb{Q})\backslash G(\mathbf{A})^1).$$

We now begin our review of the Eisenstein series by pointing out that, the first step in defining the spectral expansion of the trace formula, is to write down a formula for the kernel of the regular representation $R(f)$ in terms of the Eisenstein series. For this purpose, in what follows we shall first describe the necessary facts of the theory of Eisenstein series for $GL(r)$.

Let P be the standard parabolic subgroup associated to the partition $(r_1,\ldots,r_k)$. Write $P = M_P N_P$ and let $\lambda \in \mathfrak{a}^*_{P,\mathbb{C}} \cong \mathbb{C}^k$. To λ, and $x \in G(\mathbf{A})$ we associate the induced representation $I_P(\lambda) = I_P(\lambda, x)$, given by

$$\operatorname{ind}_{P(\mathbf{A})}^{G(\mathbf{A})}(Id_{N_P(\mathbf{A})} \otimes e^{\lambda(H_M(x))}) \otimes L^2_{\text{dis}}(M(\mathbb{Q})A_P(\mathbb{R})^0\backslash M_P(\mathbf{A}))).$$

We note that when λ is purely imaginary, $I_P(\lambda)$ is unitary. The representation $I_P(\lambda)$ acts on the Hilbert space

$$\mathbf{H}_P = L^2_{\text{dis}}(N_P(\mathbf{A})M_P(\mathbb{Q})A_P(\mathbb{R})^0\backslash G(\mathbf{A})).$$

If $\rho_P = \frac{1}{2}$ (sum of positive roots of (P, A_P)), then this action is given by

$$(I_P(\lambda, y)\phi)(x) = \phi(xy)e^{(\lambda+\rho_P)(H_P(xy)-(\lambda+\rho_P)(H_P(x))},$$

where $\phi \in \mathbf{H}_P$ and $x, y \in G(\mathbf{A})$. On the other hand, the regular representation R can be seen as an induced representation

$$R = \mathrm{Ind}_{G(\mathbb{Q})A_G(\mathbb{R})^0}^{G(\mathbf{A})}(Id_{G(\mathbb{Q})A_G(\mathbb{R})^0}).$$

What we now want, is to obtain intertwining operators between the induced representations R and $I_P(\lambda, x) = I_P(\lambda)$; this operator is given by the Eisenstein series (see also [10]). On the other hand, it is known that if $s \in S_k$, and if P' is a parabolic subgroup corresponding to the partition $(r_{s(1)}, \ldots, r_{s(k)})$, then $I_{P'}(s\lambda) \cong I_P(\lambda)$. To complete our presentation, we also need to determine intertwining operators between these two induced representations. These are given by the following formulas ((7.1), (7.2) respectively):

$$E(x, \phi, \lambda) = \sum_{\delta \in P(\mathbb{Q})\backslash G(\mathbb{Q})} \phi(\delta x)e^{(\lambda+\rho_P)(H_P(\delta x))}, \tag{7.1}$$

$$\phi \in \mathbf{H}_P,\ \lambda \in \mathbf{a}^*_{P,\mathbb{C}},\ x \in G(\mathbf{A}).$$

$$(M(s, \lambda)\phi) = \int_{N_{P'}(\mathbf{A})\cap\omega_s N_P(\mathbf{A})\omega_s^{-1}\backslash N_{P'}(\mathbf{A})} \phi(\omega_s^{-1}nx)e^{(\lambda+\rho_P)(H_P(\omega_s^{-1}nx)}e^{-(s\lambda+\rho_{P'})H_{P'}(\lambda)}dn \tag{7.2}$$

Formally, one can check that

$$E(x, I_P(\lambda, y)\phi, \lambda) = E(xy, \phi, \lambda),$$

$$M(s, \lambda)I_P(\lambda, y)\phi = I_{P'}(s\lambda, y)M(s, \lambda)\phi.$$

In the following, we give a lemma concerning the convergence of our intertwining operators, but first we need to define a certain $\mathbf{H}_P^0$ subspace of $\mathbf{H}_P$. Let, $\mathbf{m}(\mathbb{C})$ be the complex Lie algebra of M, let C be the center of the universal enveloping algebra of $\mathbf{m}(\mathbb{C})$. Then define $\mathbf{H}_P^0$ to be the set of all functions ϕ in $\mathbf{H}_P$ such that

(i) for any $x \in G(\mathbf{A})$ the function $m \to \phi(mx)$ of $M(\mathbf{A})$, is C – finite.

(ii) for $k \in K$, the span of the set of functions $\phi_k : x \to \phi(xk)$, of $G(\mathbf{A})$ is finite dimensional.

We now can state the lemma.

Lemma (7.1). If $\phi \in \mathbf{H}_P^0$ and $(Re(\lambda) - \rho_P)(\alpha^\vee) > 0\ \forall \alpha \in \Delta_P$, then the series $E(x, \phi, \lambda)$ and the integral $(M(s, \lambda)\phi)(x)$ converge absolutely.

Theorem (7.2). (Langlands): (a) If $\phi \in \mathbf{H}_P^0$, then $E(x, \phi, \lambda)$ and $M(s, \lambda)\phi$ can be analytically continued to all $\lambda \in \mathbf{a}^*_{P,\mathbb{C}}$. (b) if λ is purely imaginary, $E(x, \phi, \lambda)$ has no poles, and the operator $M(s, \lambda)$ is unitary. Moreover, we have the following functional equations:

(i) $E(x, M(s,\lambda)\phi, s\lambda) = E(x,\phi,\lambda)$;

(ii) $M(ts,\lambda) = M(t,s\lambda)M(s,\lambda) \quad s,t \in S_k$.

(c) The Eisenstein series characterizes the decomposition of R_{cont}. i.e.,

$$R_{\text{cont}} \cong \bigoplus_{\mathbf{P}} \left(\frac{1}{k!}\right) \int_{i\mathfrak{a}_P^*} I_P(\lambda).$$

In the following our aim is to give a decomposition of $L^2(G(\mathbb{Q})\backslash G(\mathbf{A})^1)$ in terms of the associated class of parabolic subgroups. By which we mean that two parabolic subgroups are *associated* if their corresponding partitions are equal in number (i.e., $k = k'$), and $(r'_1, \ldots, r'_k) = (r_{s(1)}, \ldots, r_{s(k)})$ for some $s \in S_k$. We denote by $\mathbf{P}$ a class of associated parabolic subgroups.

Let $\hat{L}_{\mathbf{P}}$ be the set of collections $F = \{F_P : P \in \mathbf{P}\}$ of measurable functions $F_P : i\mathfrak{a}_P^* \to \mathbf{H}_P$ such that

$$\text{(i)} \quad F_{P'}(s\lambda) = M(s,\lambda)F_P(\lambda) \quad s \in S_k,$$

$$\text{(ii)} \quad \| F \|^2 \stackrel{\text{def}}{=} \frac{1}{k!} \sum_{P \in \mathbf{P}} \left(\frac{1}{2\pi}\right)^k \int_{i\mathfrak{a}_P^*} \| F_P(\lambda) \|^2 \, d\lambda < \infty.$$

Under these conditions the map which sends F to the function

$$\frac{1}{k!} \sum_{P \in \mathbf{P}} \left(\frac{1}{2\pi}\right)^k \int_{i\mathfrak{a}_P^*} E(x, F_P(\lambda), \lambda) d\lambda, \quad x \in G(\mathbb{Q})\backslash G(\mathbf{A})^1,$$

defined for F in a dense subspace of $\hat{L}_{\mathbf{P}}$, extends to a unitary map of $\hat{L}_{\mathbf{P}}$ onto a closed $G(\mathbf{A})$ – invariant subspace $L^2_{\mathbf{P}}(G(\mathbb{Q})\backslash G(\mathbf{A})^1)$ of $L^2(G(\mathbb{Q})\backslash G(\mathbf{A})^1)$. And finally

$$L^2(G(\mathbb{Q})\backslash G(\mathbf{A})^1) = \bigoplus_{\mathbf{P}} L^2_{\mathbf{P}}(G(\mathbb{Q})\backslash G(\mathbf{A})^1).$$

The purpose of the following lemma is to start showing how the Eisenstein series arise in the spectral expansion of the trace formula.

Lemma (7.3). Let $f \in C_c^\infty(G(\mathbf{A})^1)$, and $R(f)$ be the regular representation into $L^2(G(\mathbb{Q})\backslash G(\mathbf{A})^1)$ with the kernel $K(x,y) = \sum_{\gamma \in G(\mathbb{Q})} f(x^{-1}\gamma y)$. Let B_P be the orthonormal basis of $\mathbf{H}_P$ (it is a Hilbert space). Then $R(f)$ is an integral operator with kernel

$$K(x,y) = \sum_P \frac{1}{k!} \left(\frac{1}{2\pi}\right)^k \int_{i\mathfrak{a}_P^*/i\mathfrak{a}_G^*} \sum_{\phi \in B_P} E(x, I_P(\lambda, f)\phi, \lambda) \overline{E(y,\phi,\lambda)} dx.$$

The idea of the proof is to start with the Duflo-Labesse [36] result in which f can be written as a finite sum of convolutions $f_1 * f_2$, where $f_i \in C_c^N(G(\mathbf{A})^1)$ for a sufficiently large N. Hence, $I_P(\lambda, f_1 * f_2) = I_P(\lambda, f_1) I_P(\lambda, f_2)$ (where $I_P(\lambda, f) = \int_{G(\mathbf{A})^1} f(x) I_P(\lambda, x) dx$). Then by applying the Schwartz inequality to the sum over ϕ and to the integral, one can reduce the Lemma to the case that $f_2(x) = f_1^*(x) = \overline{f_1(x^{-1})}$, and

$y = x$. But when $f = f_1 * f_1^*$, $R(f)$ is a positive operator, the convergence of the integral then follows from the estimate:

$$\sum_P \frac{1}{k!}\left(\frac{1}{2\pi}\right)^k \int_{|\lambda|\leq N} \int_{\phi \in \text{ finite subset of} B_P} E(x, I_P(\lambda, f)\phi, \lambda)\overline{E(y, \phi, \lambda)}$$
$$\leq K(x, x) = \sum_{\gamma \in G(Q)} f(x^{-1}\gamma x).$$

Since the right-hand side of this inequality is finite, the convergence follows.

7.2. Cusp forms, truncation, the trace formula

A basic problem in the theory of automorphic forms is to characterize the discrete spectrum of the regular representation. One may expect that the trace formula will eventually play a major role in the solution of this problem. The discrete spectrum is the direct sum of two summands, the cusp forms and the residual spectrum. When $GL(r)$ is considered over a local field the residual spectrum of this group is known [79], but, when the field is a global field the residual spectrum is only partially known [56], [105].

Before the proof of Selberg's conjecture by Müller [83] the trace formula was restricted to the study of cusp forms, since in that case one already knew that $R(f)$ is a trace class operator. However, we now know that $R(f)$ is a trace class operator on $L^2_{\text{disc}}(G(Q)\backslash G(\mathbf{A})^1)$. This makes possible to develop a satisfactory trace formula, both, on the space of cusp forms and on the discrete spectrum.

Denote the space of *cusp forms* by $L^2_{\text{cusp}}(G(Q)\backslash G(\mathbf{A})^1)$. By definition, it consists of all functions $\phi \in L^2(G(Q)\backslash G(\mathbf{A})^1)$ such that

$$\int_{N_P(Q)\backslash N_P(\mathbf{A})} \phi(nx)dn = 0, \quad \text{for almost all } x$$

and all $P \neq G$.

One knows that,

(a) $L^2_{\text{cusp}}(G(Q)\backslash G(\mathbf{A})^1)$ is a $G(\mathbf{A})^1$ – invariant subspace of $L^2(G(Q)\backslash G(\mathbf{A})^1)$.

(b) If $R_{\text{cusp}}(f)$ is the restriction of $R(f)$ to $L^2_{\text{cusp}}(G(Q)\backslash G(\mathbf{A})^1)$, then $R_{\text{cusp}}(f)$ is a Hilbert-Schmidt operator. From which we derive:

$$L^2_{\text{cusp}}(G(Q)\backslash G(\mathbf{A})^1) \subseteq L^2_{\text{dis}}(G(Q)\backslash G(\mathbf{A})^1).$$

To advance our understanding of the trace formula, we need to find an analogue of the formula $J_o(f)$ in terms of the Eisenstein series.

The first step is to decompose the space of cusp forms $L^2_{\text{cusp}}(G(Q)\backslash G(\mathbf{A})^1)$ into a direct sum of certain subspaces indexed by irreducible representations of $G(\mathbf{A})^1$. To do this, let $\Pi(H)$ denote the set of equivalence classes of irreducible representations of a group H. To each $\delta \in \Pi(G(\mathbf{A})^1)$ we associate the subspace $L^2_{\text{cusp}}(G(Q)\backslash G(\mathbf{A})^1)_\sigma$ which is equivalent to a sum of copies of σ, (i.e., isotospecial), where we recall that we are identifying representations by their spaces. Then

$$L^2_{\text{cusp}}(G(Q)\backslash G(\mathbf{A})^1) = \bigoplus_{\sigma \in \Pi(G(\mathbf{A})^1)} L^2_{\text{cusp}}(G(Q)\backslash G(\mathbf{A})^1)_\sigma. \tag{7.3}$$

Here we need the following definition, in order to prepare for the decomposition of the discrete space $L^2_{\text{dis}}(G(\mathbb{Q})\backslash G(\mathbf{A})^1)$ by analogy with the decomposition of the space of cusp forms (7.3).

Definition (7.4). A *cuspidal automorphic datum* is a pair (P,σ), where $\sigma \in \Pi(M_P(\mathbf{A})^1)$ such that

$$L^2_{\text{cusp}}(M_P(\mathbb{Q})\backslash M_P(\mathbf{A})^1)_\sigma \neq \{0\}.$$

Any two cuspidal automorphic data (P_1,σ) and (P_1',σ) are said to be *equivalent* (write $(P_1,\sigma) \sim (P_1',\sigma')$), if P_1' corresponds to the partition $(r_{s(1)},\ldots,r_{s(k)})$ and $\sigma'(m') = (s\sigma)(m') = \sigma(w_s^{-1}m'w_s)$, $s \in S_k$, $m' \in M_{P_1'}(\mathbf{A})$.

Let us denote by Ω, the set of equivalence classes $\chi = \{(P,\sigma)\}$ of these pairs. Given any P, and any $\chi \in \Omega$, define $\Omega_{P,\chi}$, to be the subspace of vectors $\phi \in \mathbf{H}_P$, such that for any $(P_1,\sigma) \notin \chi$, with $P_1 \subset P$, the integral

$$\int_{M_{P_1}(\mathbb{Q})\backslash M_{P_1}(\mathbf{A})^1} \int_{N_{P_1}(\mathbb{Q})\backslash N_{P_1}(\mathbf{A})} \psi(m)\phi(nmx)dn = 0,$$

for any function $\psi \in L^2_{\text{cusp}}(M_{P_1}(\mathbb{Q})\backslash M_{P_1}(\mathbf{A})^1)_\sigma$, and almost all x. This leads us to Langlands' result,

$$\mathbf{H}_P = \bigoplus_{\chi \in \Omega} \mathbf{H}_{P,\chi},$$

where the right-hand side is an orthogonal direct sum. Observe that in particular if $P = G$, $\mathbf{H}_G = L^2_{\text{dis}}(G(\mathbb{Q})\backslash G(\mathbf{A})^1)$. Hence

$$L^2_{\text{dis}}(G(\mathbb{Q})\backslash G(\mathbf{A})^1) = \bigoplus_{\chi \in \Omega} L^2_{\text{dis}}((G(\mathbb{Q})\backslash G(\mathbf{A})^1)_\chi). \tag{7.4}$$

Now we can choose $B_P = \bigcup_{\chi \in \Omega} B_{P,\chi}$ where $B_{P,\chi}$ is the orthonormal basis of $\mathbf{H}_{P,\chi}$. We thus have a certain kind of duality in the sense that for the analogue of $o \in \mathcal{O}$ we get $\chi \in \Omega$, so that

$$\sum_{o \in \mathcal{O}} K_o(x,y) = \sum_{\chi \in \Omega} K_\chi(x,y) = K(x,y),$$

where

$$K_\chi(x,y) = \sum_P \frac{1}{k!}\left(\frac{1}{2\pi}\right)^k \int_{i\mathfrak{a}_P^*/i\mathfrak{a}_G^*} \sum_{\phi \in B_{P,\chi}} E(x, I_P(\lambda,f)\phi,\lambda)\overline{E(y,\phi,\lambda)}d\lambda.$$

Now suppose that Q is parabolic, and $P \subset Q$, we can define the series

$$E_Q(x,\phi,\lambda) = \sum_{\delta \in P(\mathbb{Q})\backslash Q(\mathbb{Q})} \phi(\delta x)e^{(\lambda+\rho_P)(H_P(\delta x))},$$

and the *partial kernel function* $K_{Q,\chi}(x,y)$ by

$$\sum_{\{P: P \subset Q\}} \frac{1}{k!}\left(\frac{1}{2\pi}\right)^k \int_{i\mathfrak{a}_P^*/i\mathfrak{a}_G^*} \sum_{\phi \in B_{P,\chi}} E_Q(x, I_P(\lambda,f)\phi,\lambda)\overline{E_Q(y,\phi,\lambda)}d\lambda.$$

Then set:

$$K_Q(x,y) = \sum_{\chi \in \Omega} K_{Q,\chi}(x,y).$$

We also define the *partial spectral Arthur's kernel function* $k_\chi^T(x,f)$ by

$$k_\chi^T(x,f) = \sum_Q (-1)^{\dim(A_Q/A_G)} \sum_{\delta\in Q(\mathbf{Q})\backslash G(\mathbf{Q})} K_{Q,\chi}(\delta x,\delta x)\hat{\tau}_Q(H_Q(\delta x)-T_Q).$$

Then, we define the *spectral Arthur's kernel function* $k^T(x,f)$ by

$$k^T(x,f) = \sum_{\chi\in\Omega} k_\chi^T(x,f).$$

$$\sum_{o\in\mathcal{O}} k_o^T(x,f) = k^T(x,f) = \sum_{\chi\in\Omega} k_\chi^T(x,f).$$

We have already dealt with questions related to the conjugacy classes in $G(\mathbf{Q})$ and can now move on to ask analogous questions for the spectral kernels related to the elements $\chi\in\Omega$. These questions can be stated as follows.

(a) Show that $\int_{G(\mathbf{Q})\backslash G(\mathbf{A})^1} \sum_{\chi\in\Omega} |k_\chi^T(x,f)|dx < \infty.$

In fact, if we prove this, then we can write

$$\int_{G(\mathbf{Q})\backslash G(\mathbf{A})^1} k^T(x,f)dx = \sum_{\chi\in\Omega}\int_{G(\mathbf{Q})\backslash G(\mathbf{A})^1} k_\chi^T(x,f)dx = \sum_{\chi\in\Omega} J_\chi^T(f).$$

(b) Calculate $J_\chi^T(f) = \int_{G(\mathbf{Q})\backslash G(\mathbf{A})^1} k_\chi^T(x,f)dx$ explicitly, i.e., find an explicit formula for $J_\chi^T(f)$ in terms of $I_P(\lambda,f)$.

(c) What is the dependence of $J_\chi^T(f)$ on T?

For (a) we need to find a second way of expressing the kernel $K_\chi(x,x)$ in terms of the Eisenstein series, and this will be done by composing the kernel with a linear operator (*truncation operator*), which is defined on the space of functions of $G(\mathbf{Q})\backslash G(\mathbf{A})^1$. To define this operator we suppose that T is as before i.e., sufficiently regular and $\phi\in C(G(\mathbf{Q})\backslash G(\mathbf{A})^1)$ (the space of continuous functions on $G(\mathbf{Q})\backslash G(\mathbf{A})^1$). Now, define $(\Lambda^T\phi)(x)$ by:

$$\sum_{P\supset P_0} (-1)^{\dim(A_P/A_G)} \sum_{\delta\in P(\mathbf{Q})\backslash G(\mathbf{Q})} \int_{N_P(\mathbf{Q})\backslash N_P(\mathbf{A})} \phi(n\delta x)\hat{\tau}_P(H_P(\delta x)-T)dn.$$

Remark (7.5). (1) $k^T(x,f)\neq \Lambda^T K(x,x)$, (observe the similarities between them).

(2) The term when $P=G$ is just $\phi(x)$.

(3) If $G=GL(2)$, and $H_{P_0}(x) >> 0$ then if $P=P_0$, we get $\hat{\tau}_P(H_P(\delta x)-T)\neq 0$, and

$$(\Lambda^T\phi)(x) = \phi(x) - \int_{N_{P_0}(\mathbf{Q})\backslash N_P(\mathbf{A})} \phi(nx)dn.$$

Lemma (7.6). One has the following properties

(a) $\Lambda^T\circ\Lambda^T = \Lambda^T$

(b) $(\Lambda^T)^* = \Lambda^T$,

where we define the inner product $(\Lambda^T\phi_1, \phi_2)$ by an absolutely convergent integral

$$\int_{G(\mathbb{Q})\backslash G(\mathbb{A})^1} \sum_P (-1)^{\dim(A_P/A_G)} \sum_{\delta\in P(\mathbb{Q})\backslash G(\mathbb{Q})} \int_{N_P(\mathbb{Q})\backslash N_P(\mathbb{A})} \phi_1(n\delta x)$$

$$\hat{\tau}_P(H(\delta x) - T)\overline{\phi_2(x)}dndx.$$

(c) If $\phi(x)$ and its derivatives increase in a uniformly slow manner (i.e., if $|\phi(x)| \leq c \parallel x \parallel^q$ for some c and q), then $(\Lambda^T\phi)(x)$ rapidly decreases (i.e., for any q, the function $\parallel x \parallel^q |\phi(x)|$ is bounded on any Siegel set).

(d) Let P be a parabolic subgroup associated to the partition $(r_1, \ldots r_k)$. Write $W(P, P') = \{s \in S_k : P \text{ and } P' \text{ are associated}\}$. Then we have the following (*Langlands-Arthur*) formula

$$\int_{G(\mathbb{Q})\backslash G(\mathbb{A})^1} \Lambda^T E(x, \phi, \lambda)\overline{\Lambda^T E(x, \phi_1, \lambda_1)}dx =$$

$$\mathrm{vol}(\mathbb{A}_P^G/\mathbb{Z}(\Delta_P^\vee)) \sum_{P'} \sum_{s,s'\in W(P,P')}$$

$$\frac{(M(s,\lambda)\phi, M(s',\lambda_1)\phi_1)}{\sum_{\alpha\in\Delta_{P'}} (s\lambda + s'\bar{\lambda}_1)(\alpha^\vee)} e^{(s\lambda+s'\bar{\lambda}_1)(T)},$$

$\mathbb{Z}(\Delta_P^\vee)$ here is the lattice generated by $\Delta_P^\vee$.

Remark (7.7). (1) Details of the proof of the above facts can be found, for example, in [10].

(2) The (Langlands-Arthur) formula above has a one-dimensional analogue $\int_{-T}^{T} E(x,\lambda)\overline{E(x,\lambda)}dx$ obtained by Weyl 1915, and another obtained by Harish-Chandra for the semi-simple groups of real rank one.

At this point we present an interesting and challenging exercise.

Exercise (7.8). Prove directly the Lemma above for $SL_2(\mathbb{R})$.

Now we return to our questions in order to develop the trace formula. Let Λ^T act on x or y in the kernel $K(x,y)$, and set $\Lambda_i^T K_\chi(x,y)$ $i = 1,2$ for the truncated function with Λ^T acting on i^{th} variable. We then have the following:

Theorem (7.9) We have the following convergence results:

$$\text{(i)} \quad \sum_\chi \int_{G(\mathbb{Q})\backslash G(\mathbb{A})^1} |\Lambda_2^T K_\chi(x,x)|dx < \infty.$$

$$\text{(ii)} \quad \int_{G(\mathbb{Q})\backslash G(\mathbb{A})^1} \sum_{\chi\in\Omega} (\Lambda_2^T K_\chi(x,x) - k_\chi^T(x,f))dx$$

converges absolutely, and equals zero provided T is suitably regular (i.e., $t_1 >> t_2 >> \ldots >> t_r$ in $T = (t_1, \ldots, t_r)$) in a sense that depends only on the support of f.

The idea of the proof is to begin with

$$\int_{G(\mathbb{Q})\backslash G(\mathbb{A})^1} \Lambda_2^T K_\chi(x,x)dx = \int_{G(\mathbb{Q})\backslash G(\mathbb{A})^1} (\Lambda_2^T \circ \Lambda_2^T) K_\chi(x,x)dx$$

$$= \int_{G(\mathbb{Q})\backslash G(\mathbb{A})^1} \Lambda_1^T \Lambda_2^T K_\chi(x,x)dx.$$

Then (i) follows from the fact that Λ^T sends slowly increasing functions to rapidly decreasing functions. For (ii), first note that

$$\Lambda_2^T K_\chi(x,x) - k_\chi^T(x,f) =$$

$$\sum_P (-1)^{\dim(A_P/A_G)} \sum_{\delta \in P(\mathbb{Q})\backslash G(\mathbb{Q})} \int_{N_P(\mathbb{Q})\backslash N_P(\mathbb{A})} K_\chi(x, n\delta x)\hat{\tau}_P(H_P(\delta x) - T)dn$$

$$- \sum_P (-1)^{\dim(A_P/A_G)} \sum_{\delta \in P(\mathbb{Q})\backslash G(\mathbb{Q})} K_{P,\chi}(\delta x, \delta x)\hat{\tau}_P(H_P(\delta x) - T).$$

Observe that $K_\chi(x, n\delta x)$ is left $G(\mathbb{Q})$ – invariant in x. Let us change $K_\chi(x, n\delta x)$ to $K_\chi(\delta x, n\delta x)$, then the above difference equals

$$\sum_{P \neq G} \sum_{\delta \in P(\mathbb{Q})\backslash G(\mathbb{Q})} (-1)^{\dim(A_P/A_G)} \hat{\tau}_P(H_P(\delta x) - T) \int_{N_P(\mathbb{Q})\backslash N_P(\mathbb{A})}$$

$$K_\chi(\delta x, n\delta x)dn - K_{P,\chi}(\delta x, \delta x),$$

since the terms for $P = G$ in $\Lambda_2^T K_\chi(x,x)$ and $k_\chi^T(x,f)$ are equal.

We now proceed through the following example, but for more detail see [10].

Example (7.10). $G = GL(2)$, $P = P_0 = \left\{ \begin{pmatrix} * & * \\ 0 & * \end{pmatrix} \right\}$ we consider two cases for χ. The first case is that $\chi = (G, \sigma)$, with the unitary representation σ. Then our parabolic subgroup is G itself and, as in the above, the difference $\Lambda_2^T K_\chi(x,x) - k_\chi^T(x,f)$ is zero. The second case is that $\chi = (P, \xi_1 \otimes \xi_2)$ where ξ_i is Grosencharacter i.e., the character of the group $GL(1,\mathbb{Q})\backslash GL(1,\mathbb{A})^1$. Set $\delta x = y$, then one can check that the integral

$$\int_{N_P(\mathbb{Q})\backslash N_P(\mathbb{A})} K_\chi(\delta x, n\delta x)dn - K_{P,\chi}(\delta x, \delta x)$$

as appears in the difference $\Lambda_2^T K_\chi(x,x) - k_\chi^T(x,f)$, can be written as

$$\int_{N_P(\mathbb{Q})\backslash N_P(\mathbb{A})} K_\chi(y, ny)dn - K_{P,\chi}(y,y) =$$

$$\int_{M_P(\mathbb{Q})\backslash M_P(\mathbb{A})^1} \left\{ \int_{N_P(\mathbb{Q})\backslash N_P(\mathbb{A})} K(y, nmy)dn - K_P(y, my) \right\} (\xi_1 \otimes \xi_2)(m)dm,$$

and the difference in the bracket is given by

$$\{\cdot\} = \int_{N_P(\mathbb{Q})\backslash N_P(\mathbb{A})} \sum_{\gamma \in G(\mathbb{Q})} f(y^{-1}\gamma nmy)dn - \int_{N_P(\mathbb{A})} \sum_{\mu \in M_P(\mathbb{Q})} f(y^{-1}\mu nmy)dn.$$

Moreover, the first integral can be written as

$$\int_{N_P(\mathbf{A})} \sum_{\gamma\in G(Q)/N_P(Q)} f(y^{-1}\gamma nmy)dn.$$

Let us write the Bruhat decomposition for $GL(2,Q)$, which is equal to

$$GL(2,Q) = P(Q) \underset{\text{disjoint}}{\coprod} N_P(Q)\begin{pmatrix} 0 & 1 \\ 1 & 0 \end{pmatrix} P(Q). \tag{7.5}$$

(for the application of the Bruhat decomposition in general see [10] page 90).

From (7.5) we get

$$GL(2,Q)/N_P(Q) = M_P(Q) \cup N_P(Q)\begin{pmatrix} 0 & 1 \\ 1 & 0 \end{pmatrix} M_P(Q).$$

Therefore,

$$\{\cdot\} = \int_{N_P(\mathbf{A})} \sum_{\nu\in N_P(Q)} \sum_{\mu\in M_P(Q)} f(y^{-1}\nu\begin{pmatrix} 0 & 1 \\ 1 & 0 \end{pmatrix}\mu nmy)dn.$$

Assume the caracteristic functions $\hat{\tau}_P(H_P(y)-T) = \hat{\tau}_P(H_P(\delta x)-T) \neq 0$. Let $y = n_1\begin{pmatrix} r & 0 \\ 0 & r^{-1} \end{pmatrix} m_1k_1$ where $n_1 \in N_P(\mathbf{A})$, $m_1 \in M_P(\mathbf{A})^1$, $k_1 \in K$, $\log r > t_2 - t_1 >> 0$. Now to show that, $\Lambda_2^T K(x,x) = k_\chi^T(x,f)$, suppose that there is an x, such that the equality does not hold, this then implies

$$f\left(k_1\begin{pmatrix} 1 & 0 \\ n_1 & 1 \end{pmatrix}\begin{pmatrix} r^2 & 0 \\ 0 & r^{-2} \end{pmatrix}\begin{pmatrix} 1 & n_2 \\ 0 & 1 \end{pmatrix}\begin{pmatrix} m_1 & 0 \\ 0 & m_2 \end{pmatrix}k_2\right) \neq 0$$

for $n_1, n_2 \in \mathbf{A}$, $|m_1| = |m_2| = 1$, $m_i \in GL(1,\mathbf{A})^1$, and $k_1, k_2 \in K$. But f is of compact support, therefore

$$\begin{pmatrix} 1 & 0 \\ n_1 & 1 \end{pmatrix}\begin{pmatrix} r^2 & 0 \\ 0 & r^{-2} \end{pmatrix}\begin{pmatrix} 1 & n_2 \\ 0 & 1 \end{pmatrix}\begin{pmatrix} m_1 & 0 \\ 0 & m_2 \end{pmatrix} = \Psi \in C_f,$$

(C_f is the support of f which is assumed to be compact). Then computing the product

$$\Psi(1,0) = \begin{pmatrix} 1 & 0 \\ n_1 & 1 \end{pmatrix}\begin{pmatrix} r^2 & 0 \\ 0 & r^{-2} \end{pmatrix}\begin{pmatrix} 1 & n_2 \\ 0 & 1 \end{pmatrix}\begin{pmatrix} m_1 \\ 0 \end{pmatrix} =$$

$$= \begin{pmatrix} 1 & 0 \\ n_1 & 1 \end{pmatrix}\begin{pmatrix} r^2 m_1 \\ 0 \end{pmatrix}\begin{pmatrix} r^2 m_1 \\ r^2 m_1 n_1 \end{pmatrix},$$

where we have $|r^2 m_1| = r^2|m_1| = r^2 >> 0$. It follows that, C_f cannot be a compact set, and this is a contradiction. Hence

$$\Lambda_2^T K_\chi(x,x) = k_\chi^T(x,f) \quad \forall x, \ \forall \chi \in \Omega.$$

Remark (7.11). We have actually shown more. Besides the convergence of $\int_{G(Q)\backslash G(\mathbf{A})^1} \sum_{\chi\in\Omega} k_\chi^T(x,f)dx$, we have shown that the integral sign can be taken inside the sum.

Now define

$$J_\chi^T(f) = \int_{G(\mathbf{Q})\backslash G(\mathbf{A})^1} \Lambda_2^T K_\chi(x,x)dx = \int_{G(\mathbf{Q})\backslash G(\mathbf{A})^1} k_\chi^T(x,f)dx,$$

for a sufficiently large T depending only on the support of f. Then, since

$$\sum_{o} k_o^T(x,f) = \sum_{\chi} k_\chi^T(x,f),$$

we get the non-invariant form of *Arthur's trace formula*

$$\sum_{o\in O} J_o^T(f) = \sum_{\chi\in\Omega} J_\chi^T(f). \tag{7.6}$$

In the same way as we have done before by using the geometric function Γ, we can also show that $J_\chi^T(f)$ is a polynomial in T. And from the first formula for $J_\chi^T(f)$ we get

$$J_\chi^T(f) = \int_{G(\mathbf{Q})\backslash G(\mathbf{A})^1} \sum_P \frac{1}{k!}\left(\frac{1}{2\pi}\right)^k \int_{i\mathfrak{a}_P^*/i\mathfrak{a}_G^*}$$

$$\sum_{\phi\in B_P} E(x, I_P(\lambda,f)\phi,\lambda)\Lambda^T\overline{E(x,\phi,\lambda)}d\lambda dx.$$

Then by Torelli's theorem, and since Λ^T is a projection we obtain:

$$J_\chi^T(f) = \sum_P \frac{1}{k!}\left(\frac{1}{2\pi}\right)^k \int_{i\mathfrak{a}_P^*/i\mathfrak{a}_G^*}$$

$$\sum_{\phi\in B_P} \int_{G(\mathbf{Q})\backslash G(\mathbf{A})^1} \Lambda^T E(x, I_P(\lambda,f)\phi,\lambda)\Lambda^T\overline{E(x,\phi,\lambda)}dxd\lambda$$

(note the change of integral and sum)

$$= \sum_{P'} \sum_{s,s'\in W(P,P')}$$

$$\mathrm{vol}(\mathbf{A}_P^G \mathbb{Z}(\Delta_P^\vee))\frac{(M(s,\lambda)I_P(\lambda,f)\phi, M(s',\lambda)\phi)}{\prod_{\alpha\in\Delta_{P'}}(s\lambda - s'\lambda)(\alpha^\vee)} e^{(s\lambda-s'\lambda)(T)},$$

where this last equality actually holds when T is suitably regular.

Remark: The trace formula (7.6) was first introduced in [10].

Chapter VIII

The Invariant Trace Formula And Its Applications

In this final chapter we explain some of the basic ideas behind the invariant trace formula. In particular, we discuss the conditions under which one can derive a simple version of Arthur's trace formula. These trace formulas have shown to have important applications in arithmetic, topology, base change, and representation theory; in the last section we wish to explain some of the applications of the invariant trace formula in these areas.

8.1. The invariant trace formula for $GL(r)$

The trace formula given by the identity (7.6) of the preceding chapter is the first version of the generalized trace formula obtained by Arthur. This formula which in fact, as we have seen generalizes Selberg's trace formula, lacks a basic property, namely, the property of being invariant. Thus, to put the trace formula in application form, one needs an invariant trace formula of the type introduced above. This was finally achieved by Arthur; it is of course a more elaborated formula and its expression is based on the modification of the non-invariant trace formula (7.6). We now discuss the basic facts on this trace formula relying on the papers [12] and [13].

Definition (8.1). Let H be an unimodular locally compact topological group, and K_H a fixed maximal compact subgroup of H. One defines the *Hecke algebra* $\mathcal{H}_H = \mathcal{H}(H, K_H)$ to be the space of all continuous complex valued functions of compact supports f, such that for every $k, k' \in K_H$, $x \in H$

$$f(kxk') = f(x).$$

It is well known [90], and it is easy to prove that the space $\mathcal{H}_H$ is an algebra over $\mathbb{C}$ under the multiplication of functions defined by the convolution

$$(f_1 * f_2)(y) = \int_H f_1(yx^{-1}) f_2(x)\,dx.$$

The Hecke algebra $\mathcal{H}_H$ has a unit element. This element is the characteristic function of K_H normalized by the volume of K_H.

We now need to recall some notations.

(i) suppose that H is a topological group for which unitary representations and the tempered representations ([49], [50]) are defined. Then, denote by $\Pi_t(H)$ (resp. $\Pi_u(H)$) the set of equivalence classes of *irreducible tempered representations* (resp. *unitary representations*) of H. Moreover $\Pi_a(H)$ denote the set of equivalence classes of the irreducible admissible representations of H.

(ii) Let S be a finite set of valuations of $\mathbb{Q}$. Put

$$\mathfrak{a}_{M,S} = \{H_M(m) : m \in M(\mathbb{Q}_S)\}.$$

This set is in general a subgroup of $\mathbf{a}_M$, it equals $\mathbf{a}_M$ if it contains an Archimedean valuation, and it is a lattice in $\mathbf{a}_M$, otherwise.

(iii) For $\pi \in \prod_t(G(\mathbb{Q}_S))$, consider the representation

$$\pi_\lambda(x) = \pi(x)e^{\lambda(H_G(x))}, \quad x \in G(\mathbb{Q}_S),$$

and $\lambda \in \mathbf{a}^*_{G,\mathbb{C}}$, the complex vector space attached to the rational characters of G (i.e., the dual space). We use this representation to define the Fourier transform and the Paley-Wiener space.

Definition (8.2). Define the *Fourier transform* of f by

$$f_G(\pi, X) = \int_{i\mathbf{a}^*_{G,S}} tr(\pi_\lambda(f))e^{-\lambda(X)}d\lambda, \quad X \in \mathbf{a}_{G,S}.$$

In this way we can consider the maps f_G as functions on $\prod_t(G(\mathbb{Q}_S)) \times \mathbf{a}_{G,S}$. We then denote by $\mathbb{I}(G(\mathbb{Q}_S))$ the topological vector space of the maps f_G, and call it the *Paley-Wiener space*.

Suppose that f is in the Hecke algebra of $G(\mathbb{Q}_S)$, i.e., $f \in \mathcal{H}(G(\mathbb{Q}_S))$. Then, there is an open surjective map

$$T : f \to f_G.$$

The transpose of T is denoted by T', and it is given by:

$$T' : \mathbb{I}(G(\mathbb{Q}_S)) \to \mathcal{H}(G(\mathbb{Q}_S)).$$

Definition (8.3). A distribution θ on $\mathcal{H}(G(\mathbb{Q}_S))$ is said to be ***supported on characters*** if and only if it lies in the image of the transpose map T'.

To θ is associated a unique continuous map $\hat{\theta}$

$$\hat{\theta} : \mathbb{I}(G(\mathbb{Q}_S)) \to \mathbb{C},$$

given by:

$$\hat{\theta}(f_G) = \theta(f), \quad f \in \mathcal{H}(G(\mathbb{Q}_S)).$$

Then, one has the following theorem.

Theorem (8.4). A distribution θ on $\mathcal{H}(G(\mathbb{Q}_S))$ is supported on characters, if and only if $\theta(f) = 0$ for every function $f \in \mathcal{H}(G(\mathbb{Q}_S))$ such that $f_G = 0$.

Among the properties of the distributions which are supported on characters we single out the following.

Proposition (8.5). Let h be an element in $\mathcal{H}(G(\mathbb{Q}_S))$. Then any distribution θ : $\mathcal{H}(G(\mathbb{Q}_S)) \to \mathbb{C}$ which is supported on characters is invariant, in the sense that

$$\theta(h * f) = \theta(f * h) \qquad f \in \mathcal{H}(G(\mathbb{Q}_S)),$$

where $*$ is the convolution product.

A powerful result of Arthur's work shows that the maps and distributions that arise in the invariant trace formula are invariant in the sense above. The proof of these results, in the first place, is based on the definition of the invariant distribution I_M, which replaces J_M. The geometric expansion of the trace formula is based on $I_M(\gamma, f)$, and the spectral expansion on $I_M(\pi, X)$. In the following, we define $I_M(\gamma, f)$ for the argument $f \in \mathcal{H}(G(\mathbf{Q}_S))$.

Definition (8.6). $I_M(\gamma, f)$ is given by induction on the collection of the Levi subgroups L which contain M, by

$$I_M(\gamma, f) = J_M(\gamma, f) - \sum_{\substack{L \supset M \\ L \neq G}} \hat{I}^L_M(\gamma, \phi_L(f)),$$

where $\varphi_M : \mathcal{H}(G(\mathbf{Q}_S)) \to \mathcal{H}(M(\mathbf{Q}_S))$, and

$$\varphi_L(h * f) = \sum_{Q \in \mathcal{F}(M)} \varphi_L^{L_Q}(f * h),$$

where L_Q denotes the Levi component of the parabolic subgroup Q. Moreover, recall that $\hat{I}$ is defined as the only element which satisfies $\varphi'(\hat{I}) = I$, where φ' is the transpose distribution on $\mathcal{H}'(M(\mathbf{Q}_S))$.

In the following proposition we summarize some properties of I_M.

Proposition (8.7). I_M satisfies the following:

(i) I_M is supported on characters.

(ii) I_M has a germ expansion similar to that of J_M i.e.,

$$I_M(\gamma, f) = \lim_{a \to 1} \sum_{L \supset M} r^L_M(\gamma, a) I_L(a\gamma, f),$$

where γ is in general position and a belongs to the set of points in $A_M(\mathbf{Q}_S)$ whose centralizer in $G(\mathbf{Q}_S)$ equals $M(\mathbf{Q}_S)$.

(iii) Suppose that $y \in M(\mathbf{Q}_S)G(\mathbf{Q}) \cap K$. Then

$$I_{y^{-1}My}(y^{-1}\gamma y, f) = I_M(\gamma, f).$$

(iv) Suppose that v is an unramified finite valuation and that $f \in \mathcal{H}(G(\mathbf{Q}_v))$. Then

$$I_M(\gamma, f) = J_M(\gamma, f), \quad \gamma \in M(\mathbf{Q}_v).$$

To proceed further and give some other properties of I_M we need to define certain constants, and impose a condition on S, when it contains only non-Archimedean valuations.

For any Levi component $L \supset M$, set $\tilde{h} = \mathfrak{a}_L$. We write $\mathfrak{a}^{\tilde{h}}_M$ for the orthogonal complement of $\tilde{h}$ in $\mathfrak{a}_M$ and $\tilde{h}^G$ for the orthogonal complement of $\mathfrak{a}_G$ in $\tilde{h}$. Suppose that L_1, L_2 are Levi components containing M, for any bounded non-empty measurable subset U of $\mathfrak{a}^{L_2}_M$, and $\tilde{U}$ its image in $\tilde{h}^G$ under the natural map from $\mathfrak{a}^L_M$ to $\tilde{h}^G$, define

$$d^G_M(L_1, L_2) = \mathrm{vol}(\tilde{U})/\mathrm{vol}(U).$$

We now place a condition on the set S when it does not contain any Archimedean valuation. We assume that in this case all of its valuations divide a fixed rational prime p. In this case we say that S has a ***closure property*** or that it is p*–closed.*

(v) *Splitting property for* I_M. Suppose that S is p – closed. Write $S = S_1 \cup S_2$ with $S_1 \cap S_2 = \emptyset$. If $\gamma = \gamma_1\gamma_2$, $\gamma_i \in M(\mathbf{Q}_{S_i})$ $(i = 1,2)$. Then for any $f \in \mathcal{H}(G(\mathbf{Q}_S))$ of the form $f = f_1 f_2$, $f_i \in \mathcal{H}(G(\mathbf{Q}_{S_i}))$ $(i = 1,2)$ we have

$$I_M(\gamma, f) = \sum_{L_1, L_2 \supset M} d_M^G(L_1, L_2)\hat{I}_M^{L_1}(\gamma, f_{1,L_1})\hat{I}_M^{L_2}(\gamma_2, f_{2,L_2})$$

(vi) Another very basic property of I_M which is called ***descent property*** should also be explained. Briefly, it is as follows.

Suppose that $\gamma \in M(\mathbf{Q}_S)$ (S is p–closed). By γ^{M_1} $(M \supset M_1)$ one denotes the ***induced space*** in $M_1(\mathbf{Q}_S)$ i.e., γ^{M_1} is a finite union of $M_1(\mathbf{Q}_S)$ – orbits $M\{\gamma_i\}$ (in $M_1(\mathbf{Q}_S)$). For example, if $M_{1,\gamma} = M_\gamma$, then γ^{M_1} is just $M_1(\mathbf{Q}_S)$ – orbit of γ. If we call the distributions $I_{M_1}(\gamma_i, f)$ partial distributions, then descent means to find properties of

$$I_{M_1}(\gamma^{M_1}, f) = \sum_i I_{M_1}(\gamma_i, f),$$

in terms of the properties of partial distributions. Here we state some of these properties.

(a) Given $\gamma \in M(\mathbf{Q}_S)$, we have

$$I_{M_1}(\gamma^{M_1}, f) = \sum_{L \supset M} d_M^G(M_1, L)\hat{I}_M^L(\gamma, f_L).$$

where $f \in \mathcal{H}(G(\mathbf{Q}_S))$.

(b) Suppose that $\gamma \in M(\mathbf{Q}_S)$ is such that $M_\gamma = M_{1,\gamma}$. Then

$$I_M(\gamma, f) = \sum_{L \supset M} d_M^G(M_1, L)\hat{I}_M^L(\gamma, f_L).$$

Before writing down the trace formula, we need the definition of the corresponding distributions $I_M(\pi, X)$. These are invariant distributions which replace J_χ in the more explicit trace formula. As $I_M(\gamma, f)$ were related to the weighted orbital integrals J_M, the distributions $I_M(\pi, X)$ are related to ***weighted characters*** $J_M(\pi, X, f)$, which are distributions on $\mathcal{H}(G(\mathbf{Q}_S))$. In fact, it would be interesting to know the degree of similarity of these relations.

Definition (8.8). (a) For $X \in \mathbf{a}_{M,S}$, define

$$J_M(\pi, X, f) = \int_{\mathbf{a}^*_{M,S}} J_M(\pi_\lambda, f)e^{-\lambda(X)}d\lambda, \ \pi \in \Pi_{\mathrm{u}}(M(\mathbf{Q}_S)),$$

where $f \in \mathcal{H}(G(\mathbf{Q}_S))$. (For a general definition see [14, §7]).

(b) Similarly, inductively we define $I_M(\pi, X, f)$ by

$$I_M(\pi, X, f) = J_M(\pi, X, f) - \sum_{\substack{L \supset M \\ L \neq G}} \hat{I}_M^L(\pi, X, \phi_L(f)).$$

However, this is valid for arbitrary π, and $f \in \mathcal{H}(G(Q_S))$. Here we summarize some relevant properties of $I_M(\pi, X, f)$.

Proposition (8.9). $I_M(\pi, X, f)$ satisfies the following:

(i) If $\pi \in \Pi_t(M(Q_S))$. Then

$$I_M(\pi, X, f) = \begin{cases} f_G(\pi, X) & , \quad \text{if } M = G, \\ 0 & , \quad \text{if } M \neq G. \end{cases}$$

(ii) For given $\pi \in \Pi_a(M(Q_S))$, and $X \in \mathfrak{a}_{M,S}$, we have

$$I_M(\pi, X, f) = \sum_{L \supset M} d_M^G(M, L) \hat{I}_M^L(\pi, X, f_L),$$

where $f \in \mathcal{H}(G(Q_S))$. This is the *descent property*.

(iii) Let $\pi = \pi_1 \otimes \pi_2$, where $\pi_i \in \Pi_a(M(Q_S))$ $(i = 1, 2)$, $\pi \in \Pi_a(M(Q_S))$ and $X = (X_1, X_2)$ with $X_i \in \mathfrak{a}_{M,S_i}$ $(i = 1, 2)$, and $S = S_1 \cup S_2$ $S_1 \cap S_2 = \emptyset$. Then for any function $f = f_1 f_2$, $f_i \in \mathcal{H}(G(Q_{S_i}))$ $(i = 1, 2)$ we have

$$I_M(\pi, X, f) = \sum_{L_1, L_2 \supset M} d_M^G(L_1, L_2) \hat{I}_M^{L_1}(\pi_1, X_1, f_{1,L_1}) \hat{I}_M^{L_2}(\pi_2, X_2, f_{2,L_2}).$$

This is the *splitting property* for $I_M(\pi, X, f)$.

Arthur's invariant trace formula is now given as an equality between the following summations

$$(A): \qquad \sum_M |W_0^M||W_0^G|^{-1} \sum_{\gamma \in (M(Q))_{M,S}} a^M(S, \gamma) I_M(\gamma, f) \tag{8.1}$$

$$(B): \qquad \sum_{t \geq 0} \sum_M |W_0^M||W_0^G|^{-1} \int_{\Pi(M,t)} a^M(\pi) I_M(\pi, f) d\pi \tag{8.2}$$

Due to the complexity of the definitions of ingredients $a^M(\pi)$, $\Pi(M, t)$ of the spectral expansion (8.2), we shall not discuss them here, the details are given in [12, §4]. Instead, we concentrate on deriving a trace formula for $GL(r)$ when the function $f \in \mathcal{H}(G(\mathbf{A}))$ is 2-cuspidal.

Definition (8.10). Let S be $p-$closed. A function $f \in \mathcal{H}(G(\mathbf{A}))$ which is a finite sum of functions $\prod_v f_v$, $f_v \in \mathcal{H}(G(Q_v))$ is called t *– cuspidal*, whenever there are t valuations $v_i \in S$ such that $f_{v_i,M} = 0$.

Lemma (8.11). If f is 2-cuspidal, the geometric expansion of the trace formula (8.1) is given by

$$I(f) = \sum_{\gamma \in (G(Q))_{G,S}} a^G(S, \gamma) I_G(\gamma, f).$$

Proof. Let $S = S_1 \cup S_2$ $(S_1 \cap S_2 = \emptyset)$ be p – closed, which contains v_1 and v_2 respectively. From the splitting formula for I_M we observe that the distribution on the right expansion of $I_M(\gamma, f)$ vanishes unless $L_1 = L_2 = G$. Moreover, $d_M^G(G, G) = 0$

unless $M = G$. It follows that if $M \neq G$, the distribution $I_M(\gamma, f)$ equals to 0, and the corresponding term in the geometric expansion vanishes. This proves the lemma. ■

Lemma (8.12). Suppose there is a valuation v_1, such that $I_G(\gamma_1, f_{v_1}) = 0$ for any element $\gamma_1 \in G(\mathbf{Q}_{v_1})$ which is not semisimple and $\mathbf{Q}_{v_1}$ – elliptic. Suppose also that f is cuspidal at another place v_2. Then the geometric expansion of the trace formula (8.1) is given by

$$I(f) = \sum_{\gamma \in \{G(\mathbf{Q})_{\text{ell}}\}} \text{vol}(G(\mathbf{Q})_\gamma A_{G,\infty} \backslash G(\mathbf{A})_\gamma) \int_{G(\mathbf{A})_\gamma \backslash G(\mathbf{A})} f(x^{-1}\gamma x)dx,$$

where $\{G(\mathbf{Q})_{\text{ell}}\}$ denotes the set of $G(\mathbf{Q})$ – conjugacy classes of $\mathbf{Q}$ – elliptic elements in $G(\mathbf{Q})$.

Proof: One can apply the formula of the preceding lemma, as one can see, the conditions of the lemma imply that f is 2-cuspidal at v_1 and v_2. Now, if $\gamma \in G(\mathbf{Q})$ is not $\mathbf{Q}$– elliptic, it is not $\mathbf{Q}_{v_1}$ – elliptic, and $I_G(\gamma, f) = 0$. Then the lemma follows from Theorem 8.2 of [10], and from the definition of $I_G(\gamma, f)$. ■

Example (8.13). Let $G = GL(r)$, and suppose that f is 1-cuspidal at v_1. Any element $\gamma_1 \in G(\mathbf{Q}_{v_1})$ which is not $\mathbf{Q}_{v_1}$ – elliptic, belongs to a $G(\mathbf{Q}_{v_1})$ – conjugacy class

$$\delta^{G_1}, \qquad \delta_1 \in M(\mathbf{Q}_{v_1}).$$

Hence,

$$I_G(\gamma_1, f_{v_1}) = \hat{I}_M^M(\delta_1, f_{v_1,M}) = 0.$$

Therefore the first condition of the preceding lemma is satisfied. On the orther hand, suppose that f is 1-cuspidal at another valuation v_2. Then the geometric expansion of the trace formula for $GL(r)$ can be written as the formula in the preceding lemma. This will be our final formula for the geometric expansion of the trace formula for $GL(r)$ when $f \in \mathcal{H}(G(\mathbf{A}))$ is 2-cuspidal. However, we can derive a simple expression for the spectral expansion of the trace formula (8.2) under the same hypotheses, the simplification being based on the results of [20] and [111] which show that any induced unitary representation σ_1^G where $\sigma_1 \in \Pi_u(M(\mathbf{Q}_{v_1}))$ is irreducible. Moreover, we need the following result of [68], if $L^2_{\text{dis},t}(G(\mathbf{Q})A_{G,\infty} \backslash G(\mathbf{A}))$ is the subspace of $L^2(G(\mathbf{Q})A_{G,\infty} \backslash G(\mathbf{A}))$ which decomposes under $G(\mathbf{A})$ as a direct sum of representations in $\Pi_u(G(\mathbf{A}), t)$, then

$$L^2_{\text{disc},t}(G(\mathbf{Q})A_{G,\infty} \backslash G(\mathbf{A})) = \bigoplus_{\chi} L^2_{\text{disc},\chi}(G(\mathbf{Q})A_{G,\infty} \backslash G(\mathbf{A})).$$

Let us denot by $R_{\text{disc},t}$ the restritiction of the representation R on $L^2_{\text{disc},t}(G(\mathbf{Q})A_{G,\infty} \backslash G(\mathbf{A}))$. Then the simplified version of the spectral expansion of the trace formula (8.2) is simply

$$\sum_{t \geq 0} tr(R_{\text{disc},t}(f)).$$

And finally, we can write

Theorem (8.14). Let $G = GL(r)$, and f be 2-cuspidal. Then, the trace formula of G is

$$\sum_{\gamma \in \{G(\mathbf{Q})_{\text{ell}}\}} \text{vol}(G(\mathbf{Q})_\gamma A_{G,\infty} \backslash G(\mathbf{A})_\gamma) \int_{G(\mathbf{A})_\gamma \backslash G(\mathbf{A})} f(x^{-1}\gamma x)dx$$

$$= \sum_{t \geq 0} tr(R_{\text{disc},t}(f)).$$

8.2. Applications and remarks

In chapter II we discussed some applications of Selberg's trace formula. In this section we want to mention some of the applications of Arthur's invariant trace formula, then, we state some remarks concerning the generalization of the trace formula, and some relevant works.

An application of Arthur's trace formula in arithmetic, is the proof of Weil's conjecture ([116] pag 116), by Kottwitz [58]. We shall not discuss this proof here, and we refer the reader to the paper [32] for a nice presentation of the subject.

As another application, we mention the use of the trace formula to the base change problem [20], [62]. In particular, the solution of the base change problem for $GL(r)$ implies that when $Gal(E/F)$ is nilpotent Langlands' functoriality conjecture (cf. chap I) is true. To recall the base change problem suppose that E/F is a cyclic extension of degree ℓ. Consider the semi-direct product

$$\underbrace{GL(r) \times \ldots \times (GL(r)}_{\ell \text{ copies}} \rtimes Gal(E/F).$$

Let τ be an element of $Gal(E/F)$, then we denote an element of the above semi-direct produt by $g \rtimes \tau$. Define the map θ by

$$\theta(g \rtimes \tau) = \underbrace{(g, \ldots, g)}_{\ell \text{ copies}} \rtimes \tau.$$

Then the existence of θ' as in the functoriality conjecture (cf. capter I) is called the *base change problem* for $GL(r)$. Thus from the result of Arthur and Clozel one knows that θ' exists if $Gal(E/F)$ is nilpotent. Hence Langlands' functoriality conjecture is true for this case (cf. also [33], [102]).

The third application to mention here is the use of the trace formula in the problems of topological nature. In the classical setting these problems go back to the Gauss-Bonnet theorem, and Euler-Poincaré characteristic. To explain the problem in some detail, suppose that G is a connected reductive Lie group (for exemple $SL(r, \mathbb{R})$, $SO(r, \mathbb{R})$, $Sp(r, \mathbb{R})$). Moreover, suppose that G contains a maximal compact subgroup K such that

$$\text{Rank } K = \text{ Rank } G.$$

This amounts to saying that G has *discrete series*, (cf. [49]). Now, let G be such a group, then a result of Clozel and Delorme [34] asserts that the Euler-Poincaré characteristic

$$e(G, K, \pi) = \sum_{i=0} (-1)^i \dim H^i(\mathcal{G}, K; \pi)$$

associated to the $(\mathcal{G}, K)$ cohomology $H^i(\mathcal{G}, K; \pi)$ is expressed by the trace of a pseudo-coefficient. Moreover, if G has no discrete series, $e(G, K, \pi)$ must be zero. Based on this result, Arthur in his paper [15] develops a theory of Lefschetz numbers of Hecke operators for semi-simple groups with discrete series. The final calculation is based on his invariant trace formula.

By passing, we mention that up to certain reasonable hypothesis, Arthur's result can be generalized to certain non-connected reductive groups [101], see also [102], and [103].

The last application of the trace formula which we have in mind, is the role of the trace formula in representation theory. Arthur in his paper [16] begins to explain how the trace formula can give rise to certain problems in representation theory. Then, in his papers [17], [18], Arthur brings his conjectures to an up-to-date situation. We refer to these conjectures as *Arthur program* and we will discuss them elsewhere. Observe that to understand the program of Arthur, one has to have some acquaintance with the notion of *stable trace formula* [66]. In fact, it remains as a major problem in the theory of automorphic forms, to carry Arthur's invariant trace formula into an stable trace formula; namely the problem of the stabilization of the trace formula.

We now end this book by giving a remark concerning the notion of the *local trace formula* developed by Arthur in his recent work [19]. The main problem is how to relate the local trace formula to Arthur's invariant trace formula. Equivalently, whether it is possible to describe the invariant trace formula in terms of the local trace formula. Moreover, as an application of the local trace formula we mention the works [103] and [104].

Bibliography

[1] J. Arthur, *The Selberg trace formula for groups of F−rank one*, Annals of Math. (2) 100 (1974), 326-385.

[2] ________, *Automorphic representations and Number theory*, Canad. Math. Soc. Vol. 1 (1983), 49 pags.

[3] ________, *A trace formula for reductive groups I: terms associated to classes in $G(\mathbb{Q})$*, Duke Math. J. 45(1978), 911-952.

[4] ________, *The trace formula in invariant form*, Annals of Math. 114(1981), 1-74.

[5] ________, *The characters of discrete series as orbital integrals*, Invent. Math. 32(1976), 205-261.

[6] ________, *The trace formula and Hecke operators*, in [95], 11-27.

[7] ________, *A measure on the unipotent variety*, Canad. J. Math. 37(1985), 1237-1274.

[8] ________, *Local behaviour of weighted orbital integrals*, Duke Math. J. 56(1988), 233-293.

[9] ________, *Eisenstein series and the trace formula*, Proc. Sympos. Pure Math. A.M.S. Vol. 33, Part 1 (1979), 27-61.

[10] ________, *A trace formula for reductive groups II: applications of a truncation operator*, Comp. Math. 40(1980), 87-121.

[11] ________, *On a family of distributions obtained from orbits*, Canad. J. Math. Vol. 38, Nº 1 (1986), 179-214.

[12] ________, *The invariant trace formula I: local theory*, J. Amer. Math. Soc. Vol. 1, Nº 2 (1988), 323-383.

[13] ________, *The invariant trace formula II: global theory*, J. Amer. Math. Soc. Vol. 1, Nº 3 (1988), 501-554.

[14] ________, *Intertwining operators and residues I: weighted characters*, J. Funct. Analysis, Vol. 84, Nº 1 (1989), 19-84.

[15] ________, *The L^2−Lefschetz number of Hecke operators*, Invent. Math. 97 (1989), 257-290.

[16] ________, *On some problems suggested by the trace formula*, Lie Groups Representations II, Lect. Notes Vol. 1041, Springer-Verlag, Berlin – Heidelberg – New York – Tokyo (1984), 1-49.

[17] ________, *Unipotent automorphic representations: conjectures*, Soc. Math. de France, Astérisque 171-172(1989), 13-71.

[18] ________, *Unipotent automorphic representations: global motivations*, Automorphic Forms, Shimura Varieties, and L–Functions, Vol. 1, Acad. Press (1990), 1-75.

[19] ________, *A local trace formula*, Preprint 1990, to appear.,

[20] ________ and L. Clozel, *Simple Algebras, Base Change, And The Advanced Theory Of The Trace Formula*, Annals of Math. Studies Vol. 120 (1989) Princeton Univ. Press.

[21] E. Artin, *Über eine neue Art von L–Reihen*, Abh. math. Sem. Univ. Hamburg, 3 (1923), 89-108.

[22] W. L. Baily, Jr. and A. Borel, *Compactification of arithmetic quotients of bounded symmetric domains*, Annals of Math. (2) 84 (1966), 442-528.

[23] D. Barbasch and H. Moscovici, *L^2–index and the Selberg trace formula*, J. Func. Analysis, 53 (1983), 151-201.

[24] A. Borel, *Compact Clifford-Klein forms of symmetric spaces*, Topology 2 (1963), 111-122.

[25] ________, *Introduction aux groupes arithmetiques*, Hermann Paris (1969).

[26] ________, *Some finiteness properties of adèle groups over number fields*, Publication Math. IHES, 16 (1963), 5-30.

[27] ________ and G. Harder, *Existence of discrete cocompact subgroups of reductive groups over local fields*, J. Reine Angew. Math. 298 (1978), 53-64.

[28] ________ and H. Jacquet, *Automorphic forms and automorphic representations*, Proc. Sympos. Pure Math. 33, Part 1 A.M.S. (1979), 189-202.

[29] ________ and J.–P. Serre, *Corners and arithmetic groups*, Comment. Math. Helv. 48 (1973), 436-491.

[30] J.W.S. Cassels and A. Fröhlich, *Algebraic Number Theory*, Academic Press (1967).

[31] L. Clozel, *On limit multiplicities of discrete series representations in spaces of automorphic forms*, Invent. math. 83 (1986), 265-284.

[32] ________, *Numbers de Tamagawa des groupes semi-simples (d'après Kottwitz)*, Sem Bourbaki, 41 ème annèe (1988-89), n$^{\underline{o}}$ 702, Soc. Math. de France Astèrisque 177-178 (1989), 61-82.

[33] ________, *Base change for $GL(n)$*, Proceeding of I.C.M. Berkeley, Carlifornia U.S.A. (1986), 791-797.

[34] ________ and P. Delorme, *Pseudo-coefficients et cohomologie des groupes de Lie réductifs reels*, C.R. Academic Sc. Paris, t. 300, series 1, n$^{\underline{o}}$ 12 (1985), 385-387.

[35] P. Deligne, *La conjecture de Weil I*, Publication Math. IHES, 43 (1974), 273-307.

[36] M. Duflo and J.–P. Labesse, *Sur la formule des traces de Selberg*, Ann. Sci. Eć. Norm. Sup. 4e serie, t.4, (1971), 193-284.

[37] M. Eichler, *Lectures on Modular Correspondences*, Tata Inst. of Fund. Research (1957).

[38] S. Eilenberg and N. Steenrod, *Fundations Of Algebraic Topology*, (1952) Princeton Univ. Press.

[39] D. Flath, *Decomposition of representations into tensor products*, Automorphic Forms, Representations and $L-$Functions Part 1, A.M.S. (1979), 179-183.

[40] Y.Z. Flicker, *The Trace Formula and Base Change for* $GL(3)$, Lect. Notes Vol. 927 (1982) Springer-Verlag, Berlin-Heidelberg-New York.

[41] J. Fischer, *An Approach to Selberg Trace Formula via the Selberg Zeta-Function*, Lect. Notes Vol. 1253 (1987), Springer-Verlag, Berlin-Heidelberg-New York.

[42] S. Gelbart, *Automorphic Forms On Adèle Groups*, Annals of Math. Studies Vol. 83 (1975) Princeton Univ. Press.

[43] ________ and H. Jacquet, *Forms of* $GL(2)$ *from the analytic point of view*, Automorphic Forms, Representations and $L-$Functions. Part, 1 A.M.S., 213-251.

[44] ________ and F. Shahidi, *Analytic Properties of Automorphic* $L-$*Functions*, Perspective in Math. Vol. 6 (1988), Academic Press.

[45] R. Godement, *Série de Poincaré et Spitzenformen*, Expo. 10 Séminare H. Cartan Eć. Norm. Sup. 1957/58.

[46] ___________, *Domaines foundamentaux des groupes arithmétiques*, Semi. Bourbaki 15e année 1962/63, nº 257.

[47] D. Goldfeld, *Explicit formulae as trace formulae*, in [95], 281-288.

[48] Harish-Chandra, *Automorphic Forms on Semisimple Lie Groups*, Lect. Notes Vol. 62 (1968) Springer-Verlag, Berlin-Heidelberg-New York.

[49] _____________, *Discrete series for semisimple Lie groups II*, Acta Math. 116 (1966), 1-111.

[50] _____________, *Harmonic analysis on* $p-$*adic groups*, Proc. Sympos. Pure Math. 26 A.M.S. (1973), 167-192.

[51] H. Heilbronn, *Zeta functions and* $L-$*functions*, in [30], 204-230.

[52] D. A. Hejhal, *The Selberg Trace Formula for* $PSL(2,\mathbb{R})$, *Vol. 1*, Lect. Notes Vol. 548 Springer-Verlag (1976) Berlin-Heidelberg-New York.

[53] ________, *The Selberg Trace Formula for* $PSL(2,\mathbb{R})$, *Vol 2*, Lect. Notes Vol. 1001 Springer-Verlag (1983) Berlin-Heidelberg-New York.

[54] F. Hirzebruch, *Automorphe Formen und der Satz von Riemann-Roch*, Symp. Int. Top. Alg. Univ. de Mexico (1958), 92-104.

[55] ________, *Topological Methods in Algebraic Geometry*, Grund. der Math. Wiss. Vol. 131, corrected printing of the third edition (1978), Springer-Verlag, Berlin-Heidelberg-New York.

[56] H. Jacquet, *On the residual spectrum of $GL(n)$*, Lie Group Representations II, Lect. Notes Vol. 1041, Springer-Verlag, Berlin-Heidelberg-New York-Tokyo (1984), 185-208.

[57] ________ and R.P. Langlands, *Automorphic Forms on $GL(2)$*, Lect. Notes Vol. 114 (1970), Springer-Verlag, Berlin-Heidelberg-New York.

[58] R. Kottwitz, *Tamaguwa numbers*, Annals of Math. 127 (1988), 629-646.

[59] J.-P. Labesse, *Cohomologie, $L-$groupes et fonctorialité*, Comp. Math. 55(1984), 163-184.

[60] ________, *The present state of the trace formula*, Automorphic Forms, Shimura Varieties, and $L-$functions Vol. 1, Academic Press (1990), 211-226.

[61] ________, *La formule des traces d'Arthur-Selberg*, Semi. Bourbaki 37e annèe, 1984-85, n° 636, Soc. Math. de France, Astérisque 133-134 (1986), 73-88.

[62] R.P. Langlands, *Base Change For $GL(2)$*, Annals of Math. Studies Vol. 96 (1980), Princeton Univ. Press.

[63] ________, *Euler Products*, Yale Univ. Press, New Haven 1971.

[64] ________, *Problems in the theory of automorphic forms*, Lect. Notes Vol 170, Springer-Verlag, Berlin-Heidelberg-New York (1970), 18-86.

[65] ________, *Eisenstein series, the trace formula, and the modern theory of automorphic forms*, in [95], 125-155.

[66] ________, *Les D'ebuts d'une Formule des Traces Stable*, Publ. Math. Univ. Paris VII, Vol. 13 (1983) Paris.

[67] ________, *Review of the book "The Theory of Eisenstein System"*, Bull. A.M.S. Vol. 9, N° 3 (1983), 351-361.

[68] ________, *On The functional Equations Satisfied By Eisenstein Series*, Lect. Notes Vol. 544 (1976), Springer-Verlag, Berlin-Heidelberg-New York.

[69] ________, *Modular forms and $\ell-$adic representations*, Modular Functions of One Variable II, Lect. Notes Vol. 349 Springer-Verlag, Berlin-Heidelberg-New York (1973), 361-500.

[70] ________, *The dimension of spaces of automorphic forms*, Amer. J. Math. 85 (1963), 99-125.

[71] ____________, *Shimura varieties and the Selberg trace formula*, Canad. J. Math. Vol. 29, Nº 5 (1977), 1292-1299.

[72] P.D. Lax and R.S. Phillips, *Scattering Theory For Automorphic Functions*, Annals of Math. Studies Vol. 87 (1976) Princeton Univ. Press.

[73] H. Maass, *Über eine neue Art von nichtanalytischen automorphen Functionen und die Bestimmung Dirichletscher Reihen durch Functionalgleichungen*, Math. Ann. 121 (1949), 141-183.

[74] _______, *Lectures On Modular Functions Of One Complex Variables*, Tata Inst. of Fund. Research Bombay, (1964) reviesed (1983).

[75] Y. Matsushima, *On the first Betti number of compact quotient spaces of higher-dimensional symmetric spaces*, Annals of Math. (2)75 (1962), 312-330.

[76] ____________, *A formula for Betti numbers of compact locally symmetric Riemannian manifolds*, J. Diff. Geom. 1 (1967), 99-109.

[77] ____________ and S. Murakami, *On vector bundle valued harmonic forms and automorphic forms on symmetric spaces*, Annals of Math. (2)78 (1963), 365-416.

[78] H. Minkowski, *Zur Theorie der positiven quadratischen Formen*, J. Cerelle 101 (1887), 196-202.

[79] C. Moeglin and J.–L. Waldspurger, *Le specture résiduel de $GL(n)$*, Ann. Scient. Eć. Norm. Sup. 4e serie, t. 22 (1989), 605-674.

[80] *Morning Seminar* : L. Clozel, J.–P. Labesse and R.P. Langlands, Inst. Adv. Study, Princeton (1984).

[81] G.D. Mostow, *Strong rigidity for locally symmetric spaces*, Annals of Math Studies Vol. 78 (1973) Princeton Univ. Press.

[82] ___________ and T. Tamagawa, *On the compactness of arithmetically defined homogeneous spaces*, Annals of Math. Vol. 76, Nº 3 (1962), 446-463.

[83] W. Müller, *The trace class conjecture in the theory of automorphic forms*, Annals of Math. 130 (1989), 473-529.

[84] ________, *Manifolds with cusps of rank one, Spectral theory and L^2–index theorem*, Lect. Notes Vol. 1244 (1987) Springer-Verlag, Berlin-Heidelberg-New York.

[85] S.S. Rangachri and S. Shokranian, *On the dimension of the space of cusp forms for orthogonal groups of signature $(n,2)$, $n = 3,4$*, Preprint Tata Inst. Fund. Research, Bombay 1984.

[86] M. Raghunathan, *Discrete Subgroups of Lie Groups*, Ergebnise der Math. Grenz. Vol. 68 (1972) Springer-Verlag.

[87] R. Rao, *Orbital integrals in reductive groups*, Annals of Math. 96 (1972), 505-510.

[88] I. Satake, *Linear Algebra*, Marcel Dekker (1975) New York.

[89] ________, *Algebraic Structures of Symmetric Domains*, Iwanami Shoten and Princeton Univ. Press (1980).

[90] ________, *Theory of spherical functions on reductive algebraic groups over p−adic fields*, Publ. Math. IHES, 18 (1963), 229-293.

[91] ________ and S. Ogata, *Zeta functions associated to cones and their special values*, Automorphic Forms and Geometry of Arithmetic Varieties, Adv. Studies in Pure Math. 15 (1989), 1-27.

[92] G. Savin, *Limit multiplicities of cusp forms*, Invent. Math. 95 (1989), 149-159.

[93] A. Selberg, *Harmonic analysis and discontinuous groups in weakly symmetric Riemannian spaces with applications to Dirichlet series*, J. Indian Math. Soc. Vol. 20 (1956), 47-87.

[94] ________, *Automorphic functions and integral operators*, Seminars on analytic functions, Inst. Adv. Study Princeton Vol. 2 (1957), 152-161.

[95] *Selberg's Symposium*, Number Theory, Trace Formula and Discrete Groups, Acad. Press (1989).

[96] F. Shahidi, *Langlands' functoriality conjecture*, Preprint 1990 (to appear).

[97] G. Shimura, *Introduction To Arithmetic Theory of Automorphic Functions*, Iwanami Shoten and Princeton Univ. Press (1971).

[98] S. Shokranian, *Trace formula and Topology*, Coll. Brazilian Math. Soc. (1989) IMPA-Rio de Janeiro.

[99] ____________, *Limit of traces of irreducible components of supercuspidal Hecke operators*, Atas 10ª Escola de Álg. (1990), 139-143.

[100] ____________, *Some results on the dimension of space of cusp forms on classical domains of type IV*, Thesis Univ. California Berkeley (Dec. 1982).

[101] ____________, *On the twisted L^2−Lefschetz numbers for Hecke operators*, Preprint 1991.

[102] ____________, *L^2−cohomology and Hecke operators*, Preprint 1991.

[103] ____________, *Euler-Poincaré characterístic of $GL(n) \rtimes \langle\epsilon\rangle$*, Abstract Int. Cong. Math. Kyoto 1990 Japan.

[104] ____________, *A Local Lefschetz number*, Proc. School of Algebra São Paulo 1990.

[105] B. Speh, *Unitary representations of $GL(n, \mathbb{R})$ with non-trivial $(\mathcal{G}, K)-$ cohomology*, Invent. Math. 71 (1983), 443-465.

[106] G. Springer, *Introduction to Riemann Surfaces*, Addison-Wesley Pub. (1957) U.S.A.

[107] J. Tate, *Global class field theory*, in [30], 162-203.

[108] *The Selberg Trace formula and related topics*, Proceedings of a summer research conference, Contemprory Math. A.M.S. Vol. 53 (1984).

[109] A.B. Venkov, *Spectral theory of automorphic functions, the Selberg zeta function, and some problems of analytic number theory and mathematical physics*, Russian Math. Surv. 34: 3 (1979), 79-153.

[110] ________, *Spectral Theory of Automorphic Functions*, Proc. Steklov Inst. of Math. № 4 (1982).

[111] D. Vogan, *The unitary dual of $GL(n)$ over an archimedean field*, Invent. Math. 84 (1986), 449-505.

[112] A. Weil, *Sur les "formules explicites" de la théorie des nombres premiers*, Comm Sém. Math. Univ. Lund 1952, Tome supplémentaire (1952), 252-265.

[113] _____, *Sur quelques résultats de Siegel*, Summa Brasiliensis Mathematicae 1 (1946), 21-39.

[114] _____, *L'intégration dans les groupes topologiques et ses applications*, Deuxième édition, Hermann (1979) Paris.

[115] _____, *Basic Number Theory*, Crund. Math. Wiss. Band 144, Third edition (1974), Springer-Verlag, New York-Heidelberg-Berlin.

[116] _____, *Adeles and Algebraic Groups*, Progress in Math. Vol. 23 (1982) Birkhäuser, Boston-Basel-Stuttgart.

[117] S. Zucker, *Satake compactification*, Comment. Math. Helv. 58 (1983), 313-343.

Subject Index

[G]

[H]

[I]

[J]

[K]

[L]

[M]

[N]

[O]

[P]

[Q]

[R]

[S]